THE
ERGONOMICS
EDGE

Related Titles from Van Nostrand Reinhold

THE ERGONOMICS EDGE

IMPROVING SAFETY, QUALITY, AND PRODUCTIVITY

Dan MacLeod

VAN NOSTRAND REINHOLD
I(T)P   A Division of International Thomson Publishing Inc.

New York • Albany • Bonn • Boston • Detroit • London • Madrid • Melbourne
Mexico City • Paris • San Francisco • Singapore • Tokyo • Toronto

Library of Congress Cataloging-in-Publication Data

MacLeod, Dan.
    The ergonomics edge : improving safety, quality, and productivity
/ by Dan MacLeod.
      p.   cm.
    Includes bibliographical references and index.
    ISBN 0–442–01259–4
    1. Human engineering. 2. Industrial management. I. Title.
TA166.M26 1994
620.8'2—dc20
                                                    94–9580
                                                      CIP

For two of the original backwoods "ergonomists" who taught me
the value of working hard, but in the
easiest way possible:
Donald W. MacLeod, my father, Isle, Minnesota, 1914–1969
Frederick E. Burman, my grandfather, Malmo, Minnesota,
1890–1978

CONTENTS

ACKNOWLEDGMENTS

I could not have written this book alone and I wish to publicly thank those who helped me. Bertrande Blair MacLeod, my wife, provided considerable and invaluable editing assistance. My colleagues with whom I have the privilege of working—Jim Morris, Wayne Adams, Eric Kennedy, and Christine Meier—reviewed the manuscript and made countless wise suggestions, plus provided significant original input for several sections. Richard Patten and Frank Foss contributed success stories as indicated, furnishing insights in areas where I have no direct experience. J Bowling, David Roy, Carl-Johan Torarp, and Alan Appleby gracously served as reviewers, contributing much valuable direction and feedback. I am indebted to Bob Arndt and Jack Flagler for contributing to the analysis on work organization and the history of union work rules, respectively. Tom Bray, Cynthia Saunders-Beiler, Roger and Patti Carpenter, Peter Holzmann, Michael Melnik, Frank Morris, Michelle Jacobs, Terry and Judy Bryce, and John Loomis all gave much-needed moral support unstintingly. I am particularly grateful for the artistic talents of Tom Nynas and Mary Albury-Noyes whose illustrations clarified and enlivened otherwise dry text. I would also like to thank those companies and their personnel with whom I have had the pleasure of working through the years. My experiences with them constitute the inspiration and basis of this book.

I truly appreciate everyone's efforts. *The Ergonomics Edge* could not have been written without their generosity.

INTRODUCTION

Nearly everyone has experienced the following situations:

Working with your arms overhead while looking upward for
a period of time.

Trying to understand how to operate a
piece of equipment.

QUESTIONS:

1. Where would you hurt?

_____

_____

2. How would you like to work in this
posture for long periods of time?

_____

_____

1. How frustrated would you feel?

_____

_____

2. How would you like to operate
equipment like this regularly?

_____

_____

3. What kind of losses in efficiency would occur because of this?

_____

_____

_____

_____

3. How much wasted time and how many mistakes would you make because of this?

_____

_____

_____

_____

4. In your workplace do people ever . . . . . . work in awkward postures like this (or other types of contorted postures)?

_____

_____

_____

. . . work with confusing equipment, instructions, or directions equivalent to the above?

_____

_____

_____

5. With the products you produce or services you provide, do your customers ever experience similar problems?

_____

_____

_____

_____

_____

_____

6. Would your customers look elsewhere for products or services without problems such as this?

_____

_____

_____

_____

_____

_____

7. What is the cost to your business for failing to take these issues into account?

_____

_____

_____

_____

_____

_____

_____

_____

The goal of this book is to convince you that every business strategy should include improving the user-friendliness of both the workplace and the end product.

This book addresses what managers should know about ergonomics. It provides background information on basic concepts, but is not a technical book. It offers perspectives on how to think about ergonomics issues, how to apply ergonomics as part of a business strategy, and how to understand the insights on business management that this field can provide.

Ergonomics has gained visibility in much of business and industry because of regulation and litigation. The first exposure that many managers have had to the field has been media accounts of the multimillion dollar fines levied by the Occupational Safety and Health Administration (OSHA) against companies for problems related to poor ergonomics. Or, managers learn of the exploding number of product liability and workplace negligence lawsuits, which allege defects in products or dereliction of duty for failure to implement ergonomic solutions. Yet others have the simplistic impression that ergonomics means providing "cushy" jobs or slowing down production. "We pay people to work" goes the expression, "not to take it easy."

This unfortunate turn of events gives the impression that ergonomics presents just another burden. Managers hear "ergonomics" and think "more costs and more problems." However, this is a mistake. Ergonomics is part of the solution for many issues facing business. It should be part of the nation's strategy to restore industrial competitiveness.

SAVING AND EARNING MONEY

Ergonomics can save business money. By applying the basic principles of this field, employers can reduce costs related to issues like workers' compensation, turnover, and absenteeism. Operations can be made more efficient by workplace designs that create fewer errors and product defects and cause less wasted time. Products and services can be brought to market with less development costs and incurred liabilities.

Moreover, ergonomics can earn money. Business can use the tools of this field to improve the appeal and usefulness of products and, as a result, increase sales. Managers can make the workplace more user-friendly and consequently increase innovation and efficiency, and ultimately increase revenue.

Today this message is getting lost. I want to make the case that ergonomics is good for business. Indeed, it can provide any business with a competitive edge. After reading this book, it should be hard to think of the topic as only a regulatory issue. Improving the user-friendliness of both the workplace and consumer products should be part of every business strategy.

HOT TOPIC

Ergonomics is currently a hot topic. It has been called the health and safety issue of the 1990s. In particular, a perplexing issue for many managers is cumulative trauma (the injuries to joints caused, for example, by repetitive motions or static postures). Furthermore, advertisements are appearing that tout "ergonomically correct" products, and there is a need to distinguish snake oil from high octane. To understand and make good decisions, effective managers need to know these issues.

Another purpose of this book is to demystify ergonomics. Ergonomics may sound like a highly technical field of study, and some aspects of ergonomics are, in fact, complicated. But much of ergonomics is simple common sense, perhaps even a natural human reaction.

This book will have extra meaning for managers who are responsible for administrating ergonomics programs. It is not a how-to book, but does furnish some conceptual guidance and background not readily available from other sources and provides philosophical underpinnings for understanding the field.

PLAN OF THE BOOK

This book is divided into three sections:

1. *The Basics: What Every Manager Should Know.* The first section summarizes the field and outlines what managers need to know, making the case that having a good grasp of the basic principles of ergonomics improves the functioning of any business.
2. *Management Issues and Strategies.* The middle section describes in more detail the value of the field for managers in addressing a number of specific issues: quality, productivity, regulations, human resource trends, and the troublesome topic of cumulative trauma disorders.
3. *Case Studies and Workplace Programs.* The final section offers some success stories, the results of practical improvements. It includes a guide on implementing a good workplace program.

The Basics: What Every Manager Should Know

$O_{NE}^{chapter}$ THE BUSINESS CASE FOR ERGONOMICS

This chapter first provides a brief introduction to the field of
ergonomics then outlines the ways in which ergonomics can
benefit business.

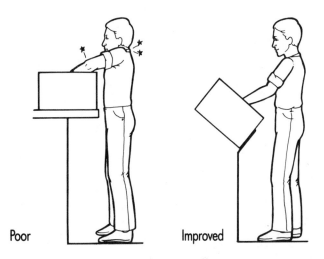

Poor   Improved

Ergonomics is all about "fitting the task to the person." Whether with a consumer
product or in the workplace, ergonomics provides opportunities for business by (1)
improving human well-being, (2) reducing costs, and (3) improving quality and
productivity.

Ergonomics is often explained by using several phrases, each of which is useful in describing the field.

WHAT IS ERGONOMICS?

Ergonomics is the field of study that seeks to design tools and tasks to be compatible with human capabilities and limitations. In this context we can think of tools and tasks broadly. A tool might range from a simple hand tool, to a written set of directions, to an entire organizational system. Similarly, a task could be either a physical or a mental activity and could be done either on or off the job. Whenever we design a more effective interrelationship between ourselves and one of our tools or tasks, that is ergonomics.

Fitting the Task to the Person

In essence, ergonomics is all about understanding human beings and human behavior—our anatomy, physiology, and psychology—and designing our tasks to fit these human requirements. We want to take conscious advantage of our unique human capabilities when we design our tools and equipment. We also want to design to counteract our human limitations—our weaknesses and frailties.

The Rules of Work

The term *ergonomics* was coined from the Greek words *ergon* (meaning "work") and *nomos* (meaning "rules"); hence, the literal definition of ergonomics is "the rules of work."

Ergonomics provides a set of conceptual guideposts for adapting workplaces, products, and services to fit human needs. The field provides a strategy for engineering design and a philosophy for good management, all with the underlying goal of improving the fit between humans and our activities. Some people have even described ergonomics as a way of thinking.

Making Tasks User-Friendly

Another good way to understand ergonomics is to use the term *user-friendly* in place of *ergonomic*. The two terms are synonymous—anything that can be described as user-friendly can also be said to be ergonomic. Things that are "unfriendly" are not ergonomic.

User-friendly means that things are easy to understand and apply, that mistakes are reduced, and that the human is treated well in the process. We have obtained the term from the designers of computer software, but the concept can be expanded into every aspect of life, whether at home or on the job. We can talk about user-friendly tools and equipment, user-friendly offices and factories, and user-friendly highway systems and shopping centers.

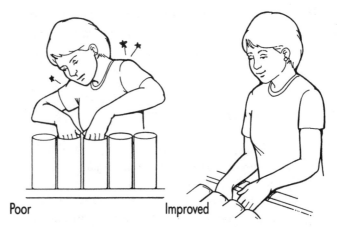

Poor   Improved

Making layout changes to permit employee to work more efficiently with less physical stress is a good example of ergonomics.

The Design Difference

Ergonomics is an interdisciplinary field that draws from anatomy, psychology, engineering, medicine, anthropology, and other related disciplines. What distinguishes ergonomics from these disciplines is the conscious goal of design—improving the fit between humans and tools.

Design Problems

Physical Issues

- long reaches
- awkward postures
- uncomfortable heights
- excessive exertion
- excessive repetition
- static postures
- contact stress

Cognitive Issues

- confusing displays or controls
- unclear presentation of information
- lack of standardization
- nonconformance to expectations
- inappropriate detail of information
- overloading mental capabilities

Applications

- tool grips to fit the contours of our hands
- furniture to fit our various heights
- tasks that do not exceed our mental or physical capabilities
- instructions and signs that are understandable
- switches and control panels that are intuitively understood
- production systems that are compatible with humans

Human–Machine Interface

Human–machine interface is a phrase that has traditionally been used to define ergonomics; however, it smacks of jargon. In its origins, this concept referred to a person working with a complex piece of equipment; for example, making sure that the person could reach all necessary controls in an airplane cockpit and that the controls themselves performed according to standard expectations. In a broader sense, we can regard ergonomics as the interaction between humans and whole systems—how people fit into entire production systems, communication networks, and decision-making processes.

All of this is ergonomics. It is a comprehensive concept that addresses the very core of work, whether for productive labor, for household chores, or even for leisure activities.

Fresh Insights through Ergonomic Glasses

One of the greatest values of ergonomics is that it causes people to think and promotes innovation. We can put on our *ergonomic glasses* and view both the workplace and the end products from a new perspective. We begin to ask questions about how a tool or production process ought to be designed to make it more human-friendly. Ultimately, that thought process can stimulate fresh insights on old problems.

Through our ergonomic glasses, we can spot problems that we have overlooked before. We can challenge assumptions, find new ways to accomplish our goals, and, sometimes, find tasks that simply do not need to be done anymore.

Ergonomics provides a healthy cross-fertilization of perspectives, with profound implications for technological and human progress. The creative mix begins with the interdisciplinary approach on which ergonomics is based, proceeds with the exchange of ideas across different industries and workplaces, and continues as experience is gained in applying ergonomic concepts to different kinds of products.

ERGONOMIC GLASSES VERSUS "THE WAY WE'VE ALWAYS DONE IT"

Once when I was doing an ergonomics survey in a manufacturing plant, I videotaped an employee struggling to dump the contents of a barrel down a chute that was about waist height. He was unaware of my presence, since I was shooting from a distance using a zoom lens.

At first he was repetitively bending over into the barrel, and removing the contents one handful at a time. Every few minutes he stopped to arch his spine backward. After a while, and a couple of attempts, he finally was able to lift the barrel up to the chute and, with a grimace, tilted it and dumped the load. Subsequently, he carried the container about ten feet back to his workbench where he started filling it up again, one handful at a time.

My intent was to show the videotape to a group of managers to help explain why they were having so many back injuries. I also wanted to start figuring out a better way to dump the barrel. I played the tape and commented on what I had observed, then played the tape over again.

"Maybe we could get a hydraulic barrel-dumper," someone said, "or a smaller container."

"Or just a lighter one," someone else contributed.

"Can we lower the chute closer to the floor?" a third asked.

I was writing down ideas on a flip-chart. An engineer said, "Look at all the time he spends on that. We could justify a barrel-dumper just on time savings, let alone workers' comp."

We continued to watch the tape in silence. Finally, the supervisor from the area said, "You know, I've walked by that job a thousand times and never really noticed it. It's stupid how that's set up."

"Why don't you just move the workbench over the chute? Skip the step of putting the stuff into the barrel?" someone asked.

"Now we're getting somewhere," I encouraged.

The supervisor responded, "That would block an aisle, but it makes me think the chute could be moved to the workbench. We could cut a little hole in the floor—the stuff just goes directly below downstairs."

The group considered a few more ideas and started to focus on making a decision. Then the senior manager in the group, silent up to now, asked, "What's he making? Isn't that the Acme product? Why can't we just make that over in Department 12, where we have that new equipment? It would cut out all of this backbreaking work entirely."

No one could think of a reason why not. There was a ripple of sheepish laughter. "That's just the way we've always done it."

I have routinely experienced this process in a variety of settings. By putting on our ergonomic glasses, we see tasks in a different light. Instead of trying to find ways of getting people to work harder, we first look for jobs that may already be too hard on people, then focus on improvements that can lead to innovation and greater efficiency. It's turning the phrase "working smarter" into a tangible process.

Most importantly, creativity occurs when end-users and designers interact. In the workplace, this can require team efforts among managers, engineers, and employees. For product development, it can mean involving customers in ergonomics studies that aid in design. In each case, we can gain insights into products and production as we tap each source of ideas, learn from each other, and spark new thoughts. End-users can be queried about their requirements and their special knowledge of the problem at hand, stimulating designers to invent better concepts.

Many of the basic principles of ergonomics are quite simple; however, by sys-

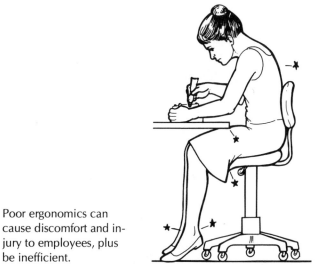

Poor ergonomics can cause discomfort and injury to employees, plus be inefficient.

tematically applying these principles to either a whole workplace or the design of a new product, ergonomics can be a tremendous source of innovation. The promise of ergonomics lies in looking at old problems from new angles, and with new methodologies. Ergonomics can characterize human requirements and provide both design criteria and techniques for evaluation.

The High Cost of Design Failures

Ergonomics is essential for business because we do not always build things the right way. Although designers sometimes think about how people fit with the tool or task, clearly at other times they do not.

All too often we plan workplaces based on "efficient movement of product" or "best locations for machines," all without much thought to how people are supposed to fit in. Too often we devise products based on the cheapest way to manufacture, or perhaps on aesthetics, but without much regard for the end-user. We think we are paying attention to the bottom line, but we may be missing important costs such as injuries, errors, and inefficient motions. Too often we expect people to adapt themselves to fit into whatever arrangement has been devised, believing that it has no associated cost. Some tasks seem to require operator training in contortionist techniques.

Unfortunately, the human body cannot adapt to everything. People have differences and they have limitations. We behave and react in certain ways that do not always fit into our standard concepts of how work should be done. If business does not understand the basic requirements of humans—how far we can reach or how we perceive information—then there will be many unnecessary costs and failures.

There can be inefficiencies in production and human error in product use. There can be frustrated employees who quit and customers who switch to a different brand. There can be injuries and workers' compensation costs and product liability suits. There may be countries whose markets are closed to U.S. products simply because we ignore the differing statures of people. The price tag for design failures can be staggering.

Every Business Can Use It

Ergonomics has broad applications that can be used in every business. The opportunities to improve the fit between people and tools surround us. Ergonomics provides business with tools to take advantage of these opportunities—creating new products, new services, new markets, and increasing sales, quality, and productivity.

Every employer can use ergonomics to improve the fit between the workplace and the people who work there. Every producer of goods and services can use ergonomics to enhance the fit between the product and the customers who use it.

Businesses are all about people; how best to use people to make products and services that best help the customers who use them. At every juncture there are

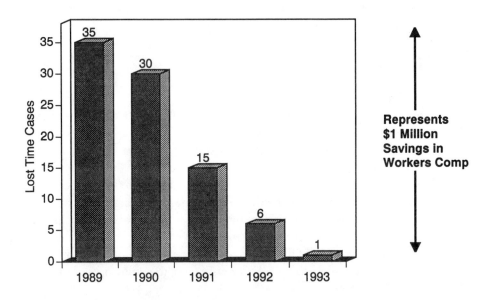

A well-implemented workplace ergonomics program can yield dramatic savings in workers' compensation costs. In this case, the company studied estimated that the cost savings were about $1 million per year in 1993. As background, in 1990, the company upgraded its safety and medical programs, plus initiated an ergonomics program. By the end of 1993, lost time cases of cumulative trauma disorders had all but been eliminated among their approximately 1000 employees.

> ## ERGONOMICS IN JAPAN
>
> A Japanese ergonomist made a study of employee suggestions that won special recognition in a large steel mill. His conclusion:
>
> *One-third of these prize-winning suggestions [in the Japanese quality circle system] can be classified as ergonomic suggestions.*
> Noro and Imada (1991)

people interacting with tools and systems. And at every point in this web of inter-actions, we can use ergonomics as guidelines for improvement, benefiting business.

This is true whether the business is a one-person firm working out of a home or a multinational corporation with an international network of plants and sales outlets. It is true for all organizations whether for profit or not; it is true for the public sector as well as private.

A New Tool for Management Problem-Solving

By systematically applying the principles and tools of ergonomics, we can cut costs and reduce these design failures. We can use the ergonomic concepts to add benefit and value for management, labor, and the consumer.

It should be self-evident that anything that makes both products and work more user-friendly is good for the bottom line. Ergonomically designed products and services lend themselves to fewer mistakes and less wasted time, and are desired more by consumers. A human-friendly workplace implies that its employees are (1) healthier, (2) better able to do their jobs, (3) have higher job satisfaction, and (4) are more productive.

Thus, ergonomics addresses an essential element of business. It provides a man-agerial and design tool to help us with the problems of today.

ERGONOMICS IN BUSINESS STRATEGY

The following are 12 ways that ergonomics can provide your business with a competitive edge.

1. People as Assets

We often hear slogans such as "People are our most important assets" or "In the end, the only source of competitive advantage is people." These statements center on truths, and hit the core of the competitive advantage of ergonomics. Ergonomics provides an approach to the design of work that is based on people. It offers both a process and a way of thinking that can turn these slogans into a concrete system of management.

2. Employee Well-Being

A good starting point for describing benefits of workplace ergonomics is simply employee well-being. Reduced injuries, fatigue, and mental stress are sufficient reasons in themselves for applying ergonomics, independent of any associated cost savings.

3. Empowerment and Involvement

Ergonomics fits well into current efforts to involve and empower people at work. The process of applying ergonomics in the workplace takes advantage of employee capabilities, ideas, and input. Many ergonomic issues can, in fact, only be addressed through the active participation of the employees who do the actual work.

Empowerment means not only allowing people to make decisions, but also providing sufficient training to increase their competency in making those decisions. For example, millions of employees and supervisors are responsible for laying out work areas and establishing work methods, and yet few have been trained in the principles and techniques of doing so correctly. Teaching people about ergonomics can help fill that gap. Moreover, being able to apply concepts of ergonomics is a skill that employees can bring to any task, and thus a valuable asset in our rapidly changing technological environment.

Incidentally, programs already established in the workplace that involve people can be used as a base for ergonomics programs. Ergonomics can take advantage of previous employee involvement efforts. Conversely, if a company has never established these formal mechanisms to involve employees, focusing on ergonomics issues is a good place to start. The concepts are relatively simple and result in direct benefit to the employees themselves, which both serves as positive reinforcement to employees for contributing ideas and provides a base for expanding to other issues.

4. Improved Morale and Employee Relations

Concern for employees and their well-being produces a payoff in improved morale. Most managers understand that good morale can be an asset, even though it is hard to estimate the dollar value. Typically, it is the little things that frustrate employees and create dissatisfaction, for example, the hard edge on a piece of equipment that the employee constantly bumps into and yet no one will fix. These are the kinds of issues that can emerge with a focus on ergonomics and can often be resolved relatively cheaply.

> ## LABOR RELATIONS TURNAROUND
>
> In my own experience in the auto industry in the 1970s, I witnessed a turnaround in relationships in many locations—from bitter adversarial ones that can only be described as "lose–lose" to relationships that fostered effective joint problem-solving and even mutual respect. The initial effort to improve these relationships was concern for worker health and safety, which began to expand into workplace ergonomics in the early 1980s. Farsighted leaders in both the UAW and the Big Three had agreed to find ways to change old patterns. Workplace safety was the common ground from which other efforts grew.
>
> This is not to suggest that unions and management *always* have the same interests; they clearly do not. However, on many issues they do share a common agenda, and working on these issues together can help to develop improved relationships on all issues. Both parties, however, must see that their common interests are being served if they hope to improve their working relationship.
>
> Industrial relations consist of a web of interactions that shift and change, sometimes daily. On some issues, interests intertwine, and on other issues or at other times, interests may conflict. Finding these moments when interests are in harmony and having ergonomic tools available to take advantage of the opportunities is a key to success.

5. Union Relations

Ergonomic issues are often good ones for joint problem-solving between management and labor. Redesigning the workplace using the principles of ergonomics is a "win–win" situation for management and labor. From the union's viewpoint, jobs are improved, injuries are decreased, people are involved and become more satisfied. These are also worthy goals from management's viewpoint, in addition to reduced costs and increased efficiencies and innovations.

Experience in many industries shows that after starting joint union–management programs on basic issues such as worker safety and workstation design, new relationships were established with positive impact on other areas. Joint programs in ergonomics can thus pave the way to other joint problem-solving efforts.

6. International Competition

As the United States and other advanced industrial nations come under increasing competition from low-wage countries, there is a simultaneous need to find new ways to compete. It is clearly possible to compete successfully against low-cost manual work by providing improved service, innovation, and quality. Taking full advantage of human capital to improve market advantage against competitors is an alternative that is not always sufficiently pursued. Ergonomics can assist in that effort, by maximizing the use of our present human resources.

7. Safety

Workplaces and products that have been designed with ergonomics in mind have a reduced risk for injury. In particular, ergonomics has proven to be effective in preventing a class of injuries known as cumulative trauma disorders, which result from damage to joints and surrounding tissue because of overuse from strenuous or highly repetitive tasks.

8. Meeting Human Resource Trends

A number of human resource trends affect the profitability of many businesses. The principles of ergonomics can be used to respond proactively to these problems and changes. Through better design of tools and equipment the costs associated with the following trends can be addressed:

- higher workers' compensation costs
- rising health care costs
- aging work force
- slower growth in the labor force
- increasing expectations of work

9. Rapidly Changing Technology

The consequences of technological change have not always been foreseen for many types of work. More than ever we need to ensure that the increasingly rapid pace of change created by new technology does not exceed human capabilities, either mentally or physically.

One of the most widespread current examples is the introduction of the personal computer into the office environment. The failure to anticipate that a typewriter could not simply be replaced by a computer has contributed to a host of employee ailments.

Mental requirements have also been affected. Sometimes jobs have evolved to the point of requiring extremely high, if not excessive, mental demands. Examples are air traffic controllers and hospital operating room technicians. Other tasks have been altered from performing manual work to monitoring a machine that does the work, resulting in extreme monotony for the employee.

Finally, we need to be reminded that despite all of our technological advances, heavy manual work still exists in many workplaces. For all of the above types of tasks, ergonomics is needed and can be effective, perhaps to an extent greater than ever before.

10. Speed of Learning

With increasingly complex equipment, training time for employees has become more extensive. Understanding instructions in some cases is now a major chore in itself. One such example is computer software, which, although supposedly written to save on human effort, can require use of thick (and sometimes infuriatingly unintelligible) manuals. In addition, the maintenance manuals for some high-tech machines used in manufacturing can be thousands of pages long and need to be computerized to permit simple use.

Fortunately, equipment that has been subjected to ergonomics evaluation can be much easier to run, and much quicker to learn to use. Training time and consequent costs can be reduced. Instruction sets themselves can also be addressed from an ergonomic perspective to make them more understandable.

A good example is, again, computer programs. Some programs are difficult to learn, with complex commands that are hard to remember. Others, such as those using graphical user interfaces (for example, Windows or the Macintosh), are relatively easy to learn and quite intuitive.

11. Customer Appeal

It should be self-evident that increasing the friendliness of any product or service—improving comfort, reducing frustrations, and so forth—improves the customer appeal. A better design achieved through good ergonomics can provide a tremendous edge over the competition.

12. Productivity

All of the above add up to more efficiency. Improving the fit between humans and tools inherently means a more effective match. Good ergonomic improvements often result in better ways of performing a task.

An ergonomically designed workplace (or product) is a more productive workplace (or product). Not exceeding human capabilities does not mean reducing output or doing less. On the contrary, good design permits more output with less human effort.

This book furnishes a variety of examples of doing more with less. Additionally, there is a rapidly growing body of literature that shows how ergonomics can improve workplace productivity (e.g., Harris, 1987; Oxenburgh, 1991). Table 1.1 summarizes a number of studies concerning the productivity payoffs from ergonomics in manufacturing and the office.

TABLE 1.1 *Productivity Increases from Ergonomics*

Researcher	Improvement	Cost-Benefit or Productivity Increase	Payback Period
Brown et al. (1991)	Material handling equipment	85% productivity increase; cost-benefit ratio of 1 to 10	Not reported
Gilbert et al. (1990)	Layouts, reaches that decrease loss of time	$5000 investment	2 weeks
Wick et al. (1990)	Various workstation changes	36% labor savings	Not reported
Steele et al. (1990)	Various workstation changes	32% time reduction	Not reported
Thomas et al. (1989)	Various workstation changes	Projected 30–50% increases in productivity	< 1 year
Schneider and Mitchell (1989)	Changed type of switch; eliminated poor neck posture	$145 cost yielded $100,000 savings per year	About 3 hours
Webb (1989)	Various workstation changes	$5000 cost; 100% productivity increase	< 3 months
Rawling and O'Halloran (1988)	Manual material handling	10–20% productivity increase	Not reported
Spilling et al. (1986)	Various workstation changes	An investment (in Norwegian crowns) of 350,000 produced savings of 3,000,000 over a 12-year period	Not reported
Dainoff (1990)	Experiment I—furniture and lighting	23.3%	Not reported
	Experiment II—furniture only	17.5%	Not reported
Francis and Dressel (1990)	Furniture	20.6%	10.8 months
Thompson (1990)	Exercise breaks	25%	Not reported

TABLE 1.1 *(continued)*

Researcher	Improvement	Productivity Increase	Payback Period
Sullivan (1990)	Furniture and organization	64.2% 32% (employees' estimate of effect of furniture) 10–15% (managers' estimate of effect of furniture)	24 months 60 months
Springer (1986)	Furniture user-adjustable seating only	15% 4–6%	5–6 months
Westinghouse Architectural Systems Division (1982)	Furniture	5.5%	23 months
American Productivity Center (1982)	Furniture	Not reported	6–24 months

USING ERGONOMICS TO CUT COSTS

There are a variety of costs associated with inadequate design, both in the workplace and in the use of products and services. Many of these costs can threaten the viability of a company. Ergonomics can help reduce these costs and make an organization more competitive.

Workplace

Employers often assume that many costs in the workplace are merely a part of doing business and beyond their control. In fact, many of these costs can be controlled—through effective ergonomics applications.

Production Barriers and Inefficiencies

A poor fit between humans and workplace systems can be inefficient in a variety of ways. Inadequate tools and equipment can be unwieldy and slow. Unclear control panels can cause confusion and impede activity. Jobs that are hard on people are often bottlenecks in production.

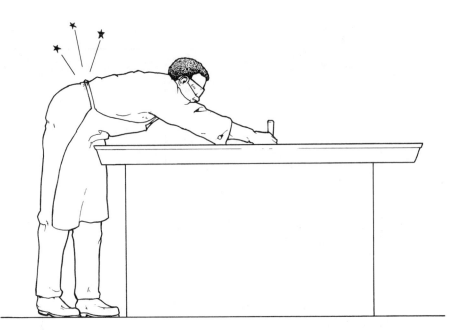

(A) Designed for failure. Costs for this job setup were high for two important reasons: (1) workers' compensation costs for back injuries as the result of working in this posture for long periods of time and (2) product defects—many of these parts were rejects and had to be redone, simply because employees could not reach the work they were required to perform and were in no position to do their jobs right the first time.

Errors and Product Defects

People working at awkward and uncomfortable workstations or with poorly planned procedures are not in a position to do their jobs correctly. Errors are more common and rework may be needed to prevent a substandard product from reaching the customer. Furthermore, unfriendly instructions or indicators, or even excessive fatigue, can contribute to costly mistakes.

Workers' Compensation

A primary direct cost-saving potential of ergonomics is the reduction of workers' compensation costs. Most workers' compensation costs (75–85 percent in many facilities that I have studied) are related to the types of injuries that ergonomics can help prevent—back injuries, wrist disorders, and assorted strains and sprains. Many companies that have initiated ergonomics programs have discovered that the programs can have a dramatic impact on their costs. Although there are

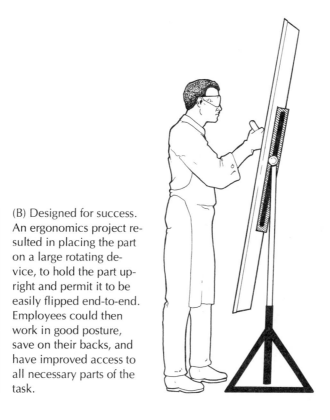

(B) Designed for success. An ergonomics project resulted in placing the part on a large rotating device, to hold the part upright and permit it to be easily flipped end-to-end. Employees could then work in good posture, save on their backs, and have improved access to all necessary parts of the task.

undoubtedly important cost issues relating to how various states administer workers' compensation plans which need attention, the need to address prevention is also critical.

Turnover and Absenteeism

One reason why employees quit or are absent is dissatisfaction resulting from unfriendly conditions: working in uncomfortable postures, being unreasonably fatigued, and experiencing symptoms of cumulative trauma. Work that hurts, or work that is frustrating, is not enjoyable and is something that people avoid if they can.

Poor Morale

Even if employees do not quit and remain on the job every day, there can be costs related to lack of ergonomic design. Poor morale may be one such cost. While the effects of poor morale are difficult to measure from an accounting point of view, it is not too difficult to accept that there can be a number of associated expenses, such as slow and inefficient work, rejection of management initiatives, or taking advantage of the employer when the opportunity presents itself.

Products and Services

The costs of providing products and services that do not adequately meet the human requirements of the customer can be many and can clearly threaten a business.

Customer Dissatisfaction

The primary cost of ignoring ergonomics in a product or service is a loss of sales. Confusing instructions, poor physical fit to human dimensions, or excessive fatigue are important reasons why the customer may seek competitive alternatives.

Product Liability

Litigation over injuries that have been related to inadequate design of products is an increasing, costly concern.

Product Development Costs

Poor attention to human issues during early design stages can cause delays in development time and increases in costs as subsequent revisions in prototypes are made to accommodate failures as the product is used.

A PROFOUND IMPACT

The crises enveloping all societies professing "scientific socialism" have distracted attention from a small, germane event in this country. It is the application of the science of ergonomics to capitalism.

George Will
Washington Post
November 16, 1989

ERGONOMICS ADDRESSES MANAGEMENT'S NEEDS

A poll* of CEOs identified their top three concerns:

1. Product quality
2. Controlling health care costs
3. Cutting production/labor costs

Ergonomics can help managers address all of these concerns:

Higher Quality

Ergonomics can be considered the other side of the quality coin. Improving workstations, tools and equipment from the human point of view can better enable employees to do their jobs right every time.

Reduced Health Care Costs

Preventing costly cumulative trauma disorders has been a prime target of ergonomics in recent years. The tools of ergonomics have a proven track record of reducing these costs substantially.

Smoother Operations and Lower Costs

The main headache for many managers is getting the product out the door while meeting customer requirements. Systematic application of the principles of ergonomics can result in lower costs and smoother operations all around, making the life of managers much easier.

Industry Week 11/19/90

Accounting Issues

There are a variety of accounting issues related to these costs. The first issue is allocating costs that are known for the enterprise as a whole back to the sources of those costs within the organization. The second issue is that many of the true costs are hidden and extra effort is needed to calculate or estimate the costs.

Some of the costs described above are fairly easily established for most businesses, for example, workers' compensation or liability costs. The problem with these cases is correctly allocating costs to specific areas of production. In a manufacturing setting, for example, supervisors or process engineers may consider these costs to be part of corporate overhead and not of real concern to them. Fortunately, many companies are beginning to charge these costs back to individual departments and a truer picture of production costs is gained at the level where many decisions are made.

Many more costs are hidden. Current accounting systems do not handle these human-related issues well. In a typical profit and loss statement, for example, there are no line items for inefficiency, poor morale, or excessive learning cycles. The costs are incurred, but they are accounted for in other, more obscure ways. Some companies are beginning to do an excellent job of estimating true costs for problem areas. Sometimes these studies occur as part of a quality improvement initiative in an attempt to estimate the "cost of failure."

Estimating the current hidden costs is important because the cost of investing in new equipment or other modifications is usually well known and sometimes seemingly expensive. Once these hidden costs are known, it is often much easier to justify equipment modifications. On occasion, the magnitude of these hidden costs can be stunning, and improvements that ordinarily would be considered quite expensive are easily justified financially.

Moreover, in the past most justifications for the purchase of capital equipment were based on a few variables such as reducing labor. Given ergonomic considerations, the calculations become more complex and more realistic.

Finally, mere recognition that these costs do occur can be helpful in making changes. Even without being able to obtain hard cost data, reflecting on these issues can often show that improvements in ergonomics can save money.

NOTES ON TERMINOLOGY

Ergonomics—Not a User-Friendly Term

The cobbler's kids have no shoes. The term *ergonomics* itself is not a user-friendly term, and in some ways is an unfortunate one. There is a certain ring to the term that may appeal to our fondness for buzzwords. However, the term is unfamiliar and often makes the field sound more complex (and expensive) than need be.

One confusion is the use of *ergonomic* versus *ergonomics*. *Ergonomic* is an adjective and *ergonomics* is a noun. When trying to decide which is which, use the following rule. If the word can be replaced with the terms *productive* or *user-friendly,* then it is *ergonomic.* If it can be replaced with the terms *productivity* or *user-friendliness,* then the term is *ergonomics.*

For example, the correct phrase is not "consult an ergonomic expert." Rather, it is "consult an ergonomics expert" (unless the person happens to also be quite productive or user-friendly). The distinction is difficult because often the grammatically correct usage does not sound right.

Incidentally, someone who specializes in er.go.*nom'*.ics is called an er.*gon'*.o.mist. Pronunciation is tricky until you hear it or say it a few times.

Human Factors and Ergonomics

These two terms are synonymous, although in the past within the United States there has been a tendency to refer to cognitive design issues as "human factors" and physical design issues as "ergonomics." In Europe the term *human factors* is generally not used and *ergonomics* has traditionally been defined to include both physical and cognitive issues.

In the United States considerable debate has taken place within the profession on this terminology. Sometimes making a firm distinction between the two branches is useful, but in the main, the two terms should be synonymous. To think otherwise is to split the human being into a traditional "body" versus "mind" dichotomy, which has definite disadvantages. From a design standpoint—for both products and work processes—both physical and cognitive issues must be taken into account anyway.

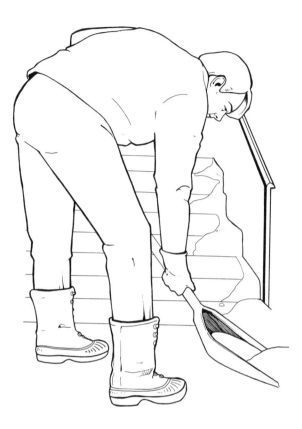

The conventional way to shovel snow.

CHAPTER ONE EXERCISE: "20 OPTIONS"

Exercise:

Part I: What is wrong with this picture? Although the illustration of shoveling snow is "normal," the task requirements are not good. List as many issues as you can.

1. _____

2. _____

3. _____

4. _____

5. _____

Part II: Study the illustration for a few minutes and write down every possible idea that comes to mind that might make this task easier. Several of the ideas must be "harebrained answers" in order to stimulate your thinking.

1. _____	11. _____
2. _____	12. _____
3. _____	13. _____
4. _____	14. _____
5. _____	15. _____
6. _____	16. _____
7. _____	17. _____
8. _____	18. _____
9. _____	19. _____
10. _____	20. _____

Answers

Part I

1. *Awkward posture: severely bent back*

2. *Repetitive, forceful motions*

3. *High overall exertion*

4. *Poor environment: cold, wet, slippery*

5. *Not satisfying task (for most people)*

In fact, many people suffer back injuries from overexertion while shoveling snow or from the severe compression, torsion, and shear forces placed on the spine while working in this posture. Furthermore, other people have experienced heart attacks from the high levels of exertion and metabolic load while shoveling.

Part II

1. *Use snowblower*	3. *Move south*
2. *Use push shovel*	4. *Have kids do it*

5. *Put canopy over steps*	**13.** *Use grating/let snow drop*
6. *Use salt*	**14.** *Use pull-type shovel*
7. *Heat the steps*	**15.** *Use flamethrower*
8. *Use lighter scoop*	**16.** *Use leaf blower*
9. *Wait until spring*	**17.** *Tramp down the snow*
10. *Use broom*	**18.** *Use a ramp/tractor & plow*
11. *Use longer-handled shovel*	**19.** *Use grating*
12. *Build house without steps*	**20.** *Use a two-handled shovel*

One of the points of this exercise is that usually, there is no one "ergonomic" solution to a problem. Often, there are quite a number of ways to make improvements. The best choice depends on the particular circumstances of each case.

There are, in fact, over 50 different ways to improve this task. Individuals

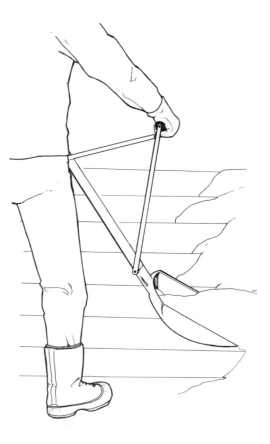

A second handle attached to the shovel provides a number of advantages.

usually can identify only a few ideas on their own. As a group, however, it is easy to come up with at least 20 options for improvement.

Some ideas are harebrained but worth writing down, since they may stimulate other ideas. For example, "flamethrowers" have been successfully used to remove snow, such as on airport runways.

The best engineering solutions are putting a canopy overhead (which keeps the snow from reaching the steps) or heating the steps. If shoveling these steps were a task at work done eight hours per day, these engineering improvements would be cost-effective. However, at home, they would often not be feasible.

However, the answer I like best (and actually use myself) is the two-handled shovel. The illustration on the opposite page shows a simple extra handle that can be attached to any shovel (unfortunately, not my invention). The two-handled shovel provides three important advantages:

1. The extra handle allows the shoveler to stand upright with the lower back in good posture and thus at less risk of injury.
2. The extra handle provides considerable mechanical advantage to lift more snow and throw it farther. For example, as winter progresses in Minneapolis, a large snowbank begins to accumulate along my driveway. With a traditional shovel, by the end of a typical winter, when I throw a load of snow on the snowbank, half of it rolls back down and has to be shoveled again. With this extra handle I can throw the snow over the snowbank (sometimes all the way onto my neighbor's driveway).
3. The extra handle provides more control over the shovel and allows improved ability to manipulate the scoop. Thus, it is easier to get into the corners of the steps and remove all of the snow.

Thus, the two-handled shovel simultaneously improves (a) safety, (b) productivity, and (c) product quality—quite a bargain. But the bottom line of this story is really low cost. I purchased the second handle in a hardware store for $12.00. Moreover, it could easily be built at a home workbench for practically nothing.

Not Just Another Snow Job

What holds us back is our lack of imagination. We can think of ways to improve the shoveling of steps, but we have a difficult time thinking of adding a second handle on a snow shovel, since we have a preconceived notion about what snow shovels are supposed to look like.

In every workplace, typically there are dozens of equivalent instances where improvements can be made for the cost of odd bits of wood and metal. But we have difficulties seeing the possibilities. We become so used to seeing the routine that we cannot imagine the options.

To break through these barriers:

- Put on ergonomic glasses.
- Brainstorm, using a team process.
- Identify the ergonomic principle that addresses the problem, then think through what kind of changes would be necessary to achieve that principle. (For example, in this instance, what does it take to permit the shoveler to straighten out the lower back?)

T~chapter~wO   PRINCIPLES OF ERGONOMICS

This chapter discusses basic principles of ergonomics. The intent here is not to provide an exhaustive treatment of the whole field or to review all of the literature or specific recommendations on how to design. Rather, the goal is to give a flavor of some of the main design concepts of ergonomics and to provide a summary of what managers should know. For further reading, consult the texts recommended in the reference list for Chapter Two.

THE HUMAN FACTOR

There are three basic human factors that must be accounted for in the design of products and workplaces: (1) people are different, (2) people have limitations, and (3) people have certain expectations and predictable responses to given situations. If we ignore these factors, the consequences are costly, both financially and in terms of human discomfort and performance.

People Are Different

We all know that people come in different shapes and sizes. Some of us are tall, others short; some are young, some old. Yet, we often set up the workplace for "one

A tall person may stoop over to work at a task, while a short person may work with shoulders hunched and elbows high to do the same task. The tall employee is a candidate for back problems and the short employee, shoulder problems. Sometimes, neither may be able to do their jobs meeting quality and production requirements because of the inadequacy of their fit to the job.

size fits all." The result is that one size fits only a small portion of the population. Large segments of the work force are working in suboptimal conditions.

Older workers may not be as robust, agile, or keen-sighted as younger workers. We rely on older people because of their experience and capabilities, but we seldom take their special needs into account when thinking about design.

Few of us are completely able-bodied in every aspect of our lives. Many of us have certain disabilities. Neglecting to consider these differences in people means that otherwise capable people might not be allowed to work at all. This policy is shortsighted not only for those challenged individuals but for employers as well because of the potential for legal liabilities under the Americans with Disabilities Act (ADA). By accommodating these differences, people with disabilities can be permitted to make their fullest contribution. Often, these workplace modifications can be achieved at relatively minor costs.

People Have Limitations

The physical and mental limits that confront humans are numerous. On the most basic level, there are overall limitations based on our stature. For example,

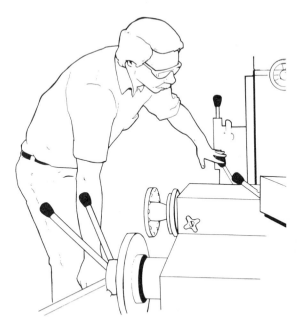

The standard industrial lathe is a classic example of failure to design for human limitations. The employee can easily suffer a sore back, neck, and shoulders from working in this posture. Furthermore, the operator is often unable to reach controls easily, particularly when trying to see the part being turned.

despite individual differences, there are limits to how far we can easily reach. Nonetheless, equipment and products are often designed as though we had infinitely long arms, with switches and controls located where we cannot reach.

The human body does not tolerate working in contorted postures for a long time, such as with arms outstretched or the body in a stooped position. Further, the wrist cannot tolerate excessive motions without injury, nor the lower back excessive lifting, bending, or twisting. Pain results and the quality of our work performance drops.

Our mental capabilities—reacting to and processing information—can also become overloaded. There is a limit to how much mental activity we can perform simultaneously or for an extended period of time. Mistakes, errors, and incorrect decisions result when we exceed these limits.

In some cases, we do not know exactly where these limitations fall, at least for any given individual. But, we do know there are limits.

Machines and physical equipment can be designed in any number of ways. But, for better or worse, human design is fixed. There is not much we can do to change the underlying shape and capabilities of the human. When there is a mismatch between humans and equipment, our recourse is to change the design of the equipment.

People Have Predictable Reactions

Throughout our lives, we have learned to associate certain actions with certain signals. We stop at red lights. We flip switches up to turn on lights. By taking into account these predictable human reactions, we can improve the design of the workplace and the operation of machines and equipment can be made more user-friendly. When we violate these expectations—for example, by orienting light switches sideways—mistakes can easily result.

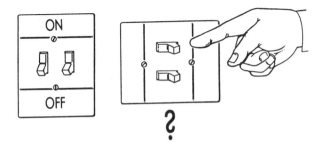

We expect lights to go on when we flip switches up (at least in the United States; other countries have different conventions). But when the switches are positioned sideways, we become confused and make errors.

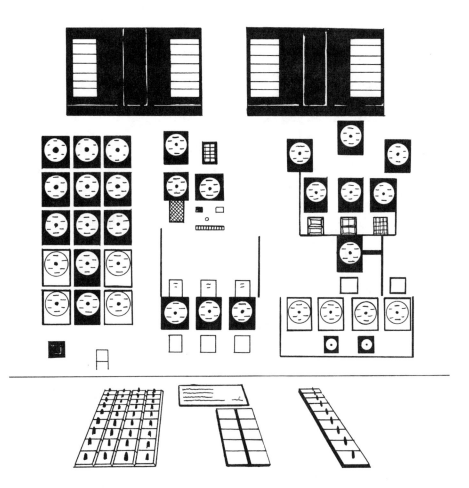

Turning on the wrong light may not result in a significant error, but in this illustration of a control panel for a nuclear power plant, flipping the wrong switch could just ruin the whole day.

In addition to our expectations concerning switches and control buttons, people respond in other predictable ways. The cognitive processes—how we think, make decisions, and react—can also be predicted. As a final example, depending on the task being performed, we can be predictably stressed, bored, or stimulated. And the lessons learned by studying these predictable responses can be integrated into good design.

40,000 Years of Ergonomic Progress

We have all had some previous experience with ergonomics. The underlying basis of the field is, in fact, more or less intuitive. We tend to arrange things to fit our own convenience, at least when we can. We keep things we need within easy reach, we change our posture when we are tired of being in the same position, and we shift to avoid glare. We try to modify our surroundings to make things easier for ourselves.

As a species, we have been doing this for a long time. In a sense, ergonomics has been a part of human progress for the past 40,000 years. Ergonomics has existed ever since the first human picked up a stone to use as a tool, capitalizing on a human capability and overcoming a human weakness. It is fair to say that the history of ergonomics parallels the entire course of human history—the continual effort of humans to recognize our capabilities and limitations and to develop tools and systems to overcome our shortcomings and build on our strengths.

So, in some ways, ergonomics is nothing new. What is new, however, is a scientific and systematic approach to designing for people. Many analytic techniques have been devised by modern-day ergonomists to permit a tangible process of design. What now lies before us is to take this natural tendency to modify our surroundings for our benefit and turn it into a conscious approach to management and product development.

To Err Is Human

What we typically think of as being human error may often in fact be poor ergonomics. Design problems can be either physical, such as locating a switch in an inaccessible position, or mental ones, such as writing a misleading set of directions. Many mistakes and errors that we make may really be attributable to poor design, an often overlooked issue that has great implications for the quality improvement process in industry.

With careful study of the errors we make, it is possible to predict human reactions, then design tools and systems to take these reactions into account. With good ergonomics, errors can be reduced in products and processes, ranging from simple household appliances such as a cooking range to complex control panels such as for an air traffic control tower.

Cognitive errors fall into three categories:

1. *Perception Errors:* The person did not grasp the needed information for any number of reasons. For example, the signal or message was not clear, there were other distracting signals, or the person was not trained in the meaning of the information.
2. *Decision Errors:* The person did not respond to the signal or information. Perhaps other decisions also had to be made quickly and the person decided the signal was not important or a priority. Or perhaps the person judged the situation incorrectly.
3. *Action Errors:* The person reacted, but activated the wrong control or activated the correct control improperly. Perhaps the controls were not laid out or did

not operate as expected. Alternatively, perhaps the operator inadvertently activated a control even when there was no signal or decision made.

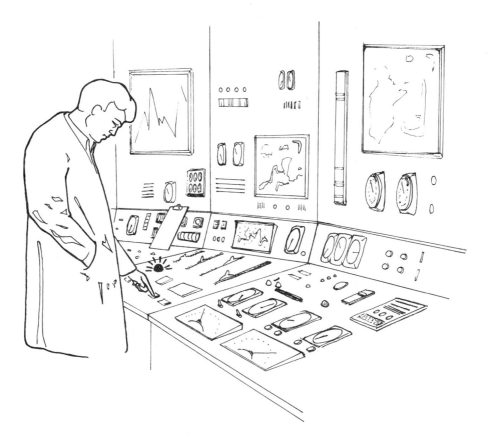

Errors can be of three kinds. (1) Perception: the operator does not notice the warning light, or the color of the light is yellow when the operator is expecting warnings to be shown in red. (2) Decision: the operator is overloaded with other decisions that simultaneously need to be made, or the operator chooses to ignore the warning, believing it is not really a problem. (3) Action: the operator reacts but activates the wrong control or moves the right control in the wrong direction.

All human factors must be taken into account when designing safe and productive workplaces. The reaches, position, and maximum loads of robotic workstations are all well-studied before being put into operation—people deserve as much attention. For guidance in accommodating these human factors we can turn to basic principles.

THE POWER OF SIMPLE PRINCIPLES

Over the past few decades a number of basic principles have emerged from the field of ergonomics. While many of these principles may appear simple, one should not underestimate the power of a few fundamental ideas applied systematically. In their basic form, these principles are the elements of "ergonomic glasses." Furthermore, in their advanced forms, more sophisticated analytic and quantitative tools can be applied.

The following sections contain two sets of principles of ergonomic design. The first ten relate to physical issues such as heights and reaches of task design, while the second ten principles relate to cognitive issues like clarity of signs and control panels.

In the following pages, you may elect to skip the details and the overview information on measurements. The basic principles are the key items.

These principles apply generally for all tasks, whether on or off the job. The terms *work* or *tasks* are used here generically to include leisure activities and household chores as well as the workplace.

The terminology here is also generic. For example, in the cognitive section, "information" can include a visual signal such as a traffic light, a dial such as the fuel gauge on a car dashboard, or a warning shout. A "control" is any device or tool that we use to execute an action, ranging from levers and switches to verbal commands.

There are certainly more than ten principles for each of the physical and cognitive aspects of ergonomics, as well as other frameworks for arranging this information. However, these allow us to address many important issues in a convenient fashion. Also, many of the principles are interrelated, but each can be used to address important problems.

In addition to these principles, there are many general rules and specific design criteria that can be generated for particular applications. However, a clear understanding of the basic principles is what is most important for managers (rather than being confronted with a long list of do's and don'ts).

As you read, ask yourself questions:

1. Do your work areas meet these principles?
2. Do your products or services meet these principles?

TEN PHYSICAL PRINCIPLES

PRINCIPLE 1—KEEP EVERYTHING IN EASY REACH
Long reaches can strain the body and make work more difficult, plus waste time. An easy way to make tasks more user-friendly is to keep frequently used items such as knobs, switches, tools, and parts within easy reach. Simple as this principle may sound, it is commonly neglected.

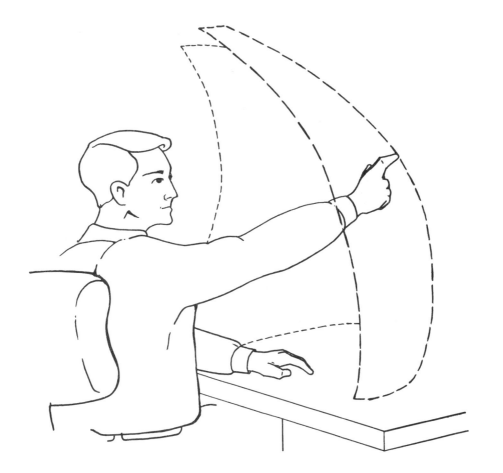

A good way to think about reaches is to visualize the hemispheric "swing space" that your arms make as they are moved.

Details

Note that this swing space is round. Yet we tend to build things rectangular, like desks and workbenches. The far corners become neglected and tend to serve as inefficient depots for clutter. A good example of designing in a hemispheric fashion are the cockpits of advanced aircraft. Pilots cannot just jump up to grab things; all controls must be within easy reach, even if it means locating switches overhead.

An important conceptual issue is to determine for whom we are really designing. Traditionally we have designed for *average* dimensions. Now the focus is on the *range* of dimensions that we need to accommodate. For example, car seats are designed to be adjusted for both taller and shorter people to reach the pedals. Emergency switches are located so that shorter people can reach them; being accessible to shorter people, they are certainly also accessible to taller people.

To make improvements, first and most important is to put on ergonomic glasses and reflect on problem reaches and the possible ways to accommodate them. Common improvements include:

- Reduce overall dimensions of the work surface. (We tend to build desks and workbenches too wide or deep.)
- Tilt the work surfaces upwards, which tends to reduce reaches.
- Make cutouts in the work surface to allow the person to work closer to needed parts and tools.
- Use lazy Susans to permit rotating necessary materials close by when needed or distant when not.

Measurements

Analytic tools are available to help guide design. The pertinent field of study is "anthropometry," the measurement of humans. Tables of measurements of characteristic human dimensions are available, as guidelines for the design of widespread applications such as kitchens, workbenches, and chairs. Computer-aided design (CAD) programs are also available that can help simulate various human dimensions and lines of sight.

PRINCIPLE 2—WORK AT PROPER HEIGHTS

A common problem is a mismatch in heights between people and the work that they are doing. This can entail working in awkward or contorted postures, which in turn can contribute to fatigue, discomfort, and even injury. Inefficiencies can also result, as people slow down to compensate for the fatigue and discomfort, or as extra work is created to overcome height differences.

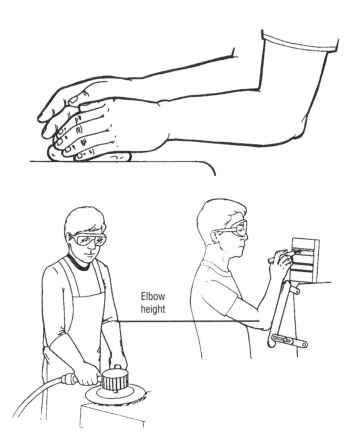

Elbow height

Generally, work should be done at elbow height, whether sitting or standing. However, heavy work should be done lower than elbow height, and lighter work, above elbow height.

Details

Conducting tasks at about elbow height helps place the body in a natural posture. However, the nature of the task also affects the proper height. Heavier work, requiring upper body strength, should be done lower than elbow height to provide better mechanical advantage. Lighter work, such as precision work or inspection tasks, should be done higher than elbow height to reduce visual distance and help improve hand–eye coordination.

The above items address the relationship between the person and the task. Additionally, there are issues of the heights within the equipment itself. Mismatches in heights in the relationship between various pieces of equipment, furniture, or flooring often cause unnecessary work (for example, lifting rather than sliding boxes). Failing to meet this simple concept has a disproportionate effect on productivity and costs because of wasted time and effort, plus it serves as a common source of frustration and irritation for the user.

Measurements

As with principles involving reach, use common sense first. Analytic tools such as anthropometry and ergonomics CAD programs equally apply. A low-tech, old-fashioned tape measure is usually the device used to measure both heights and reaches.

PRINCIPLE 3—WORK IN GOOD POSTURES

Working in awkward and contorted postures increases physical stress on the body and often reduces strength, thereby making it more difficult to do the task. Tools, furniture, and task layouts that place the body in optimal postures can increase productivity and prevent costly injuries, such as lower back problems.

The best positions in which to work are those that keep the body "in neutral." In particular, this means:

- the back with its natural "S curve" intact
- the elbows held naturally at the sides of the body
- the wrists in neutral position

Details

Note that many of these awkward positions are fine if they are only performed occasionally. The concern here is staying in a particularly bad posture for a long time.

Conversely, the body needs *some* change and mobility. Working in the same posture—even if a preferred posture—for long periods can be uncomfortable and potentially injury-provoking.

Examples of products and layouts that permit working in good posture are:

- cordless electric screwdrivers that are hinged in the middle, to permit keeping the wrists in neutral no matter what the angle of approach
- lowering the heights of elevated work surfaces, to permit relaxing the shoulders and arms
- locating frequently used objects off the floor, to eliminate the need to bend over to lift them
- using copyholders while typing or using computers, to permit working with the neck in good posture

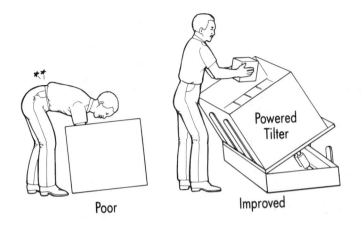

Poor

Improved

In the workplace, reaching down into tubs and bins is a common source of back injuries. Possible solutions include a hydraulic tilter as illustrated here.

A NOTE ON WRIST POSTURE

The natural posture of the wrist can be easily determined by dangling one's arms at one's sides. When completely relaxed, the wrist posture should be slightly inward and slightly forward, as it would be when holding the steering wheel of a car at locations of about ten o'clock and two o'clock.

This is what is meant by keeping the wrists "straight" or, to use a better word, "neutral." In particular, this posture is *not* the flat horizontal position required for example when playing a piano or using a computer keyboard or a mountain bike. Nor is it the erect vertical posture used when holding a small bouquet of flowers in one's fist. The natural posture is slightly inward and slightly forward.

Measurements

Postures can be measured in a number of ways. A "goniometer" is essentially an industrial-strength protractor that is used to measure angles of joints. Electronic goniometers are also coming into use, which can input data automatically into computers for sophisticated analysis. Another useful tool is combining videotape and analysis with computers, that is, watching videotapes of people at work and keying in codes for various postures. Multiple camera systems can be used to model dynamic tasks.

PRINCIPLE 4—REDUCE EXCESSIVE FORCES
Anything that can be done to minimize the exertion required to perform a task can make it more user-friendly. Needlessly excessive forces load the muscles, creating fatigue and even potential for injury.

Equipment such as tote pans or trays is often designed without adequate grips. This requires that stressful pinch grips be used. A solution as simple as providing totes with good handles can reduce needless exertion.

Details

Muscles obviously need to be exercised and used to maintain good health. Principle 4 does not suggest that we never lift a finger or become weaklings. The point is to evaluate tasks for the amount of exertion required, and make improvements. From a productivity point of view, the goal should be output (what is produced), not input (how much effort goes into it).

Common examples of correctable high exertion levels are:

1. Hand grasping forces
    *   Pinch grips are more stressful than power grips. Tools and equipment often can be improved by providing handholds that permit use of a power grip.
    *   To minimize the force needed to use hand tools, tool grips should be neither too large nor too small.

- Two-handed tools are often better than one-handed tools, since they distribute exertion, plus provide better control.
2. Arm push/pull forces
   - Use mechanical assists and power tools to minimize arm force when job requirements are elevated.
   - Fixtures that hold products, tools, and equipment can often easily reduce the effort needed to perform a task.
3. Loads on lower back
   - Design tasks so that work loads can be kept as close to the body as possible. The farther away an object is held, the harder it is on the lower back and shoulders.
   - Pushing a load is easier than pulling it. Furthermore, exertion required to push equipment such as wheeled carts can be minimized by use of larger, well-maintained wheels and good repair of the floor.

Note that when working in awkward postures, strength is reduced. Thus, designing tasks to be performed in good postures allows the work to be done with less human effort.

Measurements

Exertion can be evaluated in a variety of ways. A "force gauge" is a simple device that can be used to measure push/pull forces, as is a normal scale for lifting loads. Grasping and pinching forces can be estimated with a "hand dynamometer" and "pinch gauge," respectively.

More sophisticated measurements can also be made, primarily in a laboratory setting, although some devices are sufficiently portable and rugged to be brought to the workplace. "Force sensing resistors" can be attached either directly to the body (such as to the inside of the fingers to measure grasping force) or to the surface of equipment (such as a tool grip). "Electromyography" (EMG) involves taping electrodes on skin surface over a muscle group to detect and quantify muscle activity.

PRINCIPLE 5—MINIMIZE FATIGUE

Overloading a person's capabilities can contribute to injuries, accidents, poor quality, and lost productivity. Good task design can help prevent undesirable fatigue while maximizing efficiency.

An especially important type of fatigue is "static load," which is continuous exertion of the same muscle group over a period of time, causing discomfort and pain. By maintaining the same position for a period of time—especially when combined with high force and awkward posture—muscle groups become overloaded and blood flow reduced.

Most people have experienced static load when squirting a garden hose. After a short time the thumb muscles become overloaded and begin to hurt.

Details

Generic ways to avoid static load require nothing more than implementing some of the foregoing principles: use natural postures, change postures frequently, minimize the force required to perform tasks, and rest and stretch regularly. Specific additional examples are the following:

Static load from continuously holding an object such as a tool can be eliminated by:

- providing fixtures and jigs to support the tool and/or the product
- adding straps or handholds to the grip
- modifying handle design (for example, flanges, contour, diameter)
- using nozzles on hoses

Another type of fatigue is "metabolic load," which refers to overall exertion levels and the general weariness that can result. Body nutrients become depleted and the person becomes sluggish and less alert.

Problems with such highly strenuous tasks can be minimized in several ways. Mechanization is an obvious option, but certain administrative changes can also help. Frequent short breaks allow people to recharge their batteries and are actually more effective than longer but fewer breaks. The intensity and duration of mental and physical effort can be limited throughout the day, peak work loads spread over more time, and demanding tasks rotated with less demanding tasks. Reducing environmental extremes (see below) can also reduce fatigue.

Measurements

Static load is measured as the duration of time in which a muscle group is contracted. Heart rate monitors are often used to measure metabolic load. Electromyography and other sophisticated instruments can be used in a laboratory setting to measure subtle effects of fatigue. There are also self-rating scales such as the Borg scale.

PRINCIPLE 6—REDUCE EXCESSIVE REPETITION
The number of repeated motions required to perform a task has a profound impact on the wear and tear on the body. Excessive repetitions can create injury to sensitive tissue and joints, as well as contribute to inefficient use of time. Wherever feasible, repetitive motions should be reduced.

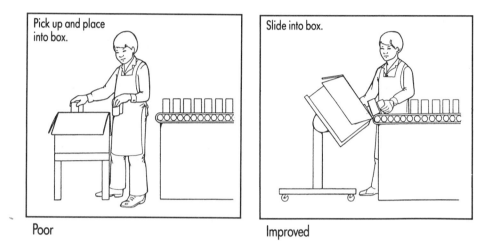

Pick up and place into box.

Slide into box.

Poor   Improved

Packing tasks often involve repetitive "pick-and-place" hand motions. Workstation layout changes can be made to eliminate many of these unnecessary hand and arm motions, specifically by allowing the product to be slid into the boxes.

Details

Several general ways to minimize excessive repetition are:

- Use machines to do repetitive tasks.
- Design for motion efficiency and use the most efficient workstation layout possible.
- Use good work methods and efficient technique.
- Eliminate double-handling.

Measurements

Repetitions are defined as movements or exertions made by a major joint or body link. They are measured simply by observing and counting, either while directly watching the task or by reviewing videotapes.

> **PRINCIPLE 7—PROVIDE CLEARANCE AND ACCESS**
> It is important that work areas be designed with enough space to both get the task done and have easy access to everything needed. Adequate clearance is needed for heads, arms, torsos, knees, and feet. There should be no obstructions between a person and the items needed to accomplish the task.

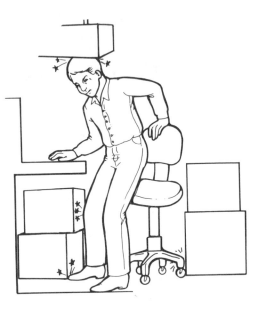

Clearance can often be evaluated simply by determining if the workstation causes people to bump into things.

Details

In general, by constructing tasks to fit larger people, it is obvious that smaller people will also have sufficient room. Access can be improved by reducing reaches and reorganizing equipment, shelves, or other barriers. Addressing the size and shape of maintenance ports is another area of improvement.

Visual access is concerned with the ability of the person to see the work at hand. It is as important to eliminate line-of-sight barriers as it is to provide physical access.

Measurements

Many of these issues can be addressed through common sense. Additionally, the anthropometric tables and tape measures mentioned previously can be used.

PRINCIPLE 8—MINIMIZE CONTACT STRESS
Direct pressure or "contact stress" is an issue common to many tasks. In addition to being uncomfortable, it can inhibit nerve function and blood flow.

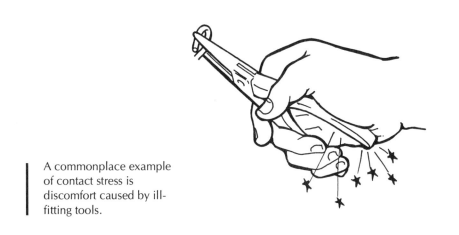

A commonplace example of contact stress is discomfort caused by ill-fitting tools.

Details

Contact stress commonly affects the palms, forearms, thighs, shins, and feet. Palm pressure can be reduced by changing the shape, contour, size, and covering of tool handles. In this way, the pressure to hold them will be more evenly distributed over the palm.

Leaning the forearms against sharp or hard edges for support or thighs rubbing against obstructions are other examples of contact stress. Improvements include padding and/or rounding the edges in question, or other modifications made.

Providing good padding and contour for chairs also applies here. A related item involves standing. The human body does not well-tolerate standing on hard surfaces for long periods of time. The rigidity is fatiguing and can contribute to lower back and leg problems. Examples of ways to provide the necessary cushioning are antifatigue floor mats or wearing cushioned insoles.

Measurements

There are no convenient analytic tools yet for these problems. An approach is simply to look for or ask employees about bruises, calluses, and indentations or red marks on the affected parts of the body.

> **PRINCIPLE 9—PROVIDE MOBILITY AND CHANGE OF POSTURE**
> There is no single "correct" posture best for an entire workday. The human body needs change and mobility. Good ergonomic design provides for opportunities to change positions, move around, or alternate between sitting and standing.

Alternate between
sitting and standing

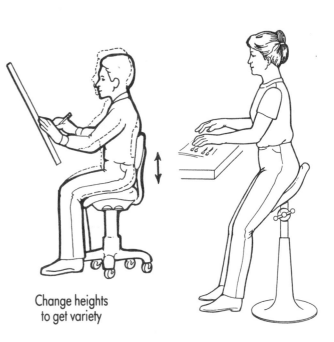

Change heights
to get variety

Many jobs can also be done using a sit–lean stand offering an alternative posture to both prolonged sitting and standing.

Details

Adjustability makes it easier to customize workstations to fit human needs. Adjustable furniture and equipment permit changes in posture, plus make it possible that better heights and reaches are maintained and pressure points and awkward postures avoided.

Some good examples of providing adjustability are:

*   good office chairs and footrests
*   automobile seats, steering wheels, and mirrors
*   barber and hairdresser chairs
*   adjustable workstations or adjustable-height platforms
*   providing sit–lean stands

PRINCIPLE 10—MAINTAIN A COMFORTABLE ENVIRONMENT
The environment in which work is performed can directly and indirectly affect not only the comfort and health of people, but also the quality and efficiency of the work being done. There are many issues here, but three important goals are:

- provide appropriate lighting
- avoid temperature extremes
- isolate vibration

Details

Appropriate Lighting

A simple but often overlooked solution is to provide appropriate lighting. The quantity and quality of light at tasks can either enhance or obscure the details of the work.

Common problems include:

- glare that shines in workers' eyes
- shadows that hide details
- poor contrast between the task and the background

Options for improvement include:

- diffusers or shields to minimize glare
- better placement of lights to avoid glare
- task lighting or indirect overhead lighting to soften shadows
- backlighting to enhance contrast

Temperature Extremes

Being too hot or cold while performing a task can cause discomfort and fatigue, and may contribute to health problems. The working temperature can be affected by increasing or decreasing the following:

- air temperature and velocity
- amount of shielding from hot or cold objects
- physical demands of the work performed
- type and amount of clothing worn
- humidity

Isolate Vibration

Working with vibrating tools and equipment can cause discomfort as well as potential injury to sensitive tissue. Vibration can damage the nerves and blood ves-

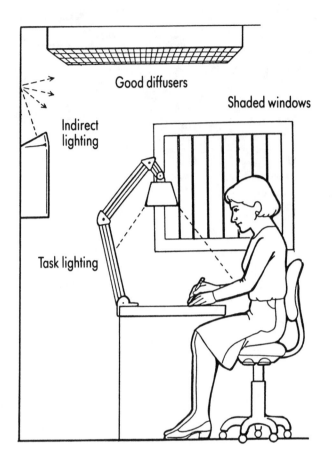

Good diffusers

Shaded windows

Indirect
lighting

Task lighting

> Good lighting that does not produce glare or shadows is an example of maintaining a comfortable environment.

sels in the hands as well as affect the discs and surrounding tissue in the lower back. Exposure to both whole-body and hand–arm vibration can be minimized, using the following example techniques:

- change equipment speeds and feeds
- perform routine maintenance
- mount equipment on vibration-dampening pads
- use cushioned floor mats for standing operations
- include vibration-dampening materials in or on tools

Other

There are many other issues in this category, such as toxic chemicals and excessive noise. These issues are often considered part of industrial hygiene.

TEN COGNITIVE PRINCIPLES

> ### PRINCIPLE 1—STANDARDIZE
> Many errors are caused because there is inconsistency in how information is displayed and how controls work. To prevent mistakes, a general rule is to insure that similar devices work the same way. Agreeing on a standard

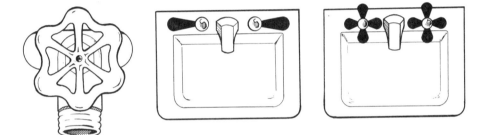

Operation of single faucets (such as outdoor spigots for garden hoses) has been fairly standardized, and even provided a popular rule for activation: "Righty-tighty, lefty-loosey." Double faucets, on the other hand, are often not standardized. How would *you* turn these faucets on? In the workplace, lack of standardization between pipe fittings, pressure valves, and other equipment has caused mistakes with serious repercussions.

Examples

In a manufacturing plant, a standard could be agreeing to wire the controls of several pieces of equipment in the same way, so that an operator can easily switch from one to another. A further example is color coding wires or pipes. A more highly formalized standard is a government regulation for labeling hazardous chemicals. Definitions are also standards—agreeing to call things in a uniform way to prevent confusion.

Sometimes it does not matter which system is adopted, as long as it is always the same. Other times, an arbitrarily set standard may conflict with the user's perceptions. In the latter case, the standard should be consistent with these human perceptions.

PRINCIPLE 2—USE STEREOTYPES
A stereotype is a commonly held expectation of what people think is supposed to to happen when they recognize a signal or activate a control. Good design should take advantage of these perceptions and expectations.

The concept of a stereotype is closely related to that of a standard, but much less consciously determined. Whereas a standard is a formal agreement to eliminate inconsistencies, a stereotype is an informal convention that has evolved through time. A good standard often follows a stereotype. Conversely, a standard that has become culturally ingrained through widespread use can become a stereotype.

Examples

- moving a lever forward to make a device go forward or go faster
- rotating a knob clockwise to make a pointer turn to the right or make it increase
- using red to mean stop, or danger

Clocks provide a good example of a device based on stereotypes that has become highly standardized. The hands on the clock face always turn in the same direction so universally that we use the term *clockwise* to describe this type of movement. Most clocks also are divided into 12 hours. While there may be arguments that the day could be divided up more conveniently into other systems, such as decimals, the point is that the current system has been established by convention and considerable confusion avoided.

In contrast, there is no accepted consensus for how to change the time setting on a clock, especially digital clocks, and much confusion results. Some people have found changing the clocks in their cars so bewildering that they give up altogether and live through summer daylight saving time an hour off.

PRINCIPLE 3—LINK ACTIONS WITH PERCEPTIONS
Ideally, there should be a strong relationship between the perception of the need to take an action and the action itself—a compatibility between a display of information and a control. Good design means configuring things so that it is self-evident what one is supposed to do.

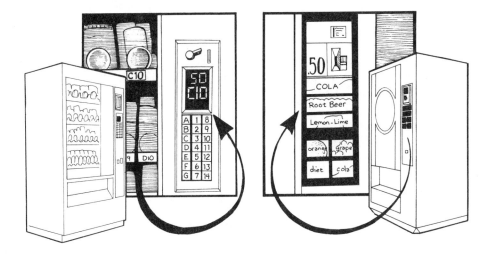

It was late in the afternoon of a tiring workday and I wanted chocolate and sugar. I was thinking about a candy bar all the way down the hall to the vending machine. I put in my money—my last 50 cents—noting that the chocolate bar I wanted was "C11," then reached to the buttons and punched "C," then "1," and then another "1." To my instant horror I realized that I should have pushed the button "11." With the anticipated taste of the chocolate still in my mouth, out of the machine tumbled a pack of breath mints. The design of this candy machine is prone to errors, since the selection buttons are distant from the choices of candy. The opportunities for mistakes are abundant: the customer can misread the number of the candy bar, forget the correct number before being able to press a button, or inadvertently press the wrong button; the buttons are arranged in a different pattern than the candy; and there are two possible ways for the customer to select the number "11" or higher (by pressing for example a "1" and then another "1" which is common on some machines or by pressing the "11" which is common on others).

In contrast, the soft drink machine strongly links actions with perceptions: the symbol of the brand and type of drink is imprinted directly on a rather large activating button. All of the possible errors of the candy machine are precluded in one simple design.

Examples

The control panel for a complex piece of machinery may have a number of dials that provide information on the various functions of the machine. The controls that govern these machine functions should be linked closely with the dials, so that it is intuitive which control affects which function and which dial.

Mnemonics also helps in this regard, such as on a car gearshift where the "R" stands for Reverse and the "P" for Park. Computer keyboard commands have also exploited this concept. A "CONTROL-P" for Printing is much easier to remember and more closely linked to the perception of the need to print than is some arbitrary numeral or other letter.

Another good example of this principle is the use of recorded verbal warnings in commercial airplane cockpits. For example, if an airplane flies too low, a recorded voice comes on commanding urgently, "Pull up! Pull up! Pull up!" This system links very quickly the perception of danger with the action of increasing altitude and is far superior to previous systems of warning lights and buzzers or of relying on the pilots to check the altimeter from time to time.

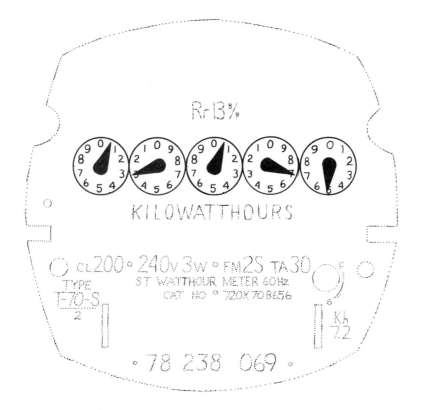

The traditional gas or water meter dials (that alternate clockwise and counterclockwise directions for each digit) are examples of complex presentation of information that is nearly impossible to read correctly without careful study and double checking. Even the professionals from the utility companies make mistakes.

PRINCIPLE 4—SIMPLIFY PRESENTATION OF INFORMATION
Sometimes too much information is provided, or it is provided in too complex a fashion. In general, good designs provide simplified displays.

Example

A common example of simplified design is the use of visual images—photographs, icons, or signs—rather than either written or spoken words.

> ### PRINCIPLE 5—PRESENT INFORMATION AT THE APPROPRIATE LEVEL OF DETAIL
>
> Careful consideration should be given as to what information the user needs to know. There are many options for the level of detail of information that is presented and the choices can either enhance or hinder performance. Sometimes users need only general information; signals should be correspondingly approximate and general. At other times, detailed and precise information is crucial.

Digital displays are often best when precise information is required. Analog gauges are faster and clearer for giving general indication. For relative information, moving pointers are often best.

Examples

A common example is a training session. Sometimes all that the participants in the training session need is an overview, while other participants in other sessions require in-depth materials. Unfortunately, in training sessions this principle is too often violated: participants are given unneeded information or information at an unwarranted level of detail.

The design of signs, instruction manuals, and control panels all can benefit from evaluation. What information does the user need to know?

In many cases icons can convey information better than other means. The message can be clear and quickly grasped, even despite language and literacy differences.

Make It Visible

First, the message must be visible. The size and location should be appropriate at the distance from which it is to be observed and there should be no obstructions. Signs and labels should contrast with their background.

Make It Distinguishable

The message should also be distinguishable from other surrounding signals and information. Multiple signals, such as with warning lights or alarms, should not be so similar that they can be confused. For example, a fire alarm should have a pattern or pitch that is distinct from a "process down" alarm.

There should be adequate space to separate messages from one another.

Make It Interpretable

Finally, the message should be interpretable. An example of one type of issue is avoiding use of characters that look alike—1l, B8, QO—and breaking up long strings—(612)854–0112. Another type of issue is matching the message with the training of the user.

> ### PRINCIPLE 7—USE REDUNDANCIES
> Sometimes, one message is insufficient. Because mistakes are easy to make and humans have many limitations, it is important to provide the same information in more than one way.

A zip code used in conjunction with a city and state is an example of redundancy that is helpful. With the zip code, the city and state really are not needed—all of the information the post office needs is contained in the code. However, it is very easy to make a mistake and write down the wrong number, and the names serve to help the post office correct the error.

Examples

A police car that uses both sirens and flashing lights is a common example; if people miss one cue, they can pick up on another.

The tradition of filling out checks with the dollar amounts written out as well as in numeric form helps to overcome bad handwriting and making mistakes. Stop signs at road intersections have three redundancies: color (red), shape (octagonal), and wording (STOP).

PRINCIPLE 8—USE PATTERNS
The human eye grasps patterns well. Information presented as a pattern can often be understood much more quickly and accurately than otherwise.

Examples

When displaying numerical data, graphs are much easier to read and interpret than columns of numbers. Bar charts are especially good for comparing numbers and line charts are good for showing trends.

In control panels for complex equipment, grouping appropriate controls can ease use. Similarly, placing controls in patterns or in a context can help indicate to the user what to do. Conversely, using patterns that differ from expectations can be confusing.

Aligning dials so that all point in the same direction when operations are normal promotes easier detection of abnormal conditions. The eye can quickly perceive the pointer that is out of alignment, and thus needing attention.

> ### *PRINCIPLE 9—PROVIDE VARIABLE STIMULI*
> Humans detect a novel stimulus more readily than a constant one, since our senses fatigue easily with continuous exposure. For example, flashing lights are easier to spot than unchanging lights. Buzzers that sound only infrequently are noticed more than recurrent ones. Thus, it is important to avoid excessive use of a single way of presenting information.

Examples

Methods of providing warnings are good examples. Written warnings are notorious for being ignored, since they quickly become part of the background and simply are not read. The detailed safety instructions on commercial airlines that are provided verbatim for every flight are seemingly not heard by experienced travelers; only when the information is provided in an unusual way (such as a video featuring an unconventional character) do the frequent fliers pay attention. Some ambulances have been designed to exploit this principle: the sirens switch patterns from time to time to evoke attention.

Applications can affect many arenas. Training sessions that switch from one format to another can be more effective than unremitting exposure to a single method of presentation.

NOT HEARING A SIREN

In a steel mill, an overhead crane routinely traversed an area carrying loads of heavy steel bars. As it moved, a siren blared a loud and constant warning to anyone walking underneath.

I was in the vicinity of the crane along with the safety director of the mill doing an ergonomics task analysis when suddenly there was a huge "bebop" noise that startled us. The crane had approached us, siren wailing as loud as ever, but we did not notice it—we had become accustomed to the sound. Only when the operator changed the pattern of the siren did we hear it, and then with great effect.

PRINCIPLE 10—PROVIDE INSTANTANEOUS FEEDBACK
An additional principle that helps prevent errors is to provide feedback to the user on the course of action taken. Furthermore, the sooner the feedback is given, the easier it is to determine if an error has been made or not.

Examples

Pilots signal that a message has been received with the word *roger* (a term itself chosen because it is more easily understood than a simple "yes"). When taking a telephone message, repeating numbers helps clarify that the correct message is being conveyed. The keys on computer keyboards are deliberately designed to click to help indicate that a character has been successfully transmitted.

The total absence of feedback prevents even knowing if a mistake has been made or not, and drastically affects the likelihood that an error will be repeated.

CHAPTER TWO EXERCISE: THE BUSINESS CASE FOR ADJUSTABILITY

1. List the types of adjustment available in cars to accommodate differing heights of people:

2. How do you feel when you drive a car that is not adjusted for you?

3. What would happen if you drove eight hours per day, every day, in a position that was not right for you?

4. What would you do if you went to a car dealer to buy a new car and the seat was not adjustable, not even forward and backward?

5. What happens when employees do not have the means to adjust workplace equipment and furniture?

6. What happens when customers do not have the means to adjust products to fit them?

ANSWERS

1. *seats: forward/backward*

   seats: up/down, tilt

   seats: reclinable seat back

   steering wheel: up/down

   all mirrors

2. *either cramped or can't reach the pedals*

   uncomfortable

3. *fatigued*

   cranky

4. *probably not buy the car*

   look elsewhere

5. *undue fatigue & job dissatisfaction*

   errors, inefficiencies, poor quality

6. *probably not buy the product*

   look elsewhere

chapter THREE

BUSINESS PERSPECTIVES ON ERGONOMICS

This chapter suggests ways to think about the application of ergonomics. Most of this was written with the workplace in mind, but the concepts can easily be transferred to consumer products and other applications of ergonomics.

PERSPECTIVES

It Does Not Have to Be Complicated

In almost every aspect of ergonomics, we can address the issues on two levels: (1) as a subject for sophisticated science and (2) as a matter of common sense. Both are equally important, but depend on the circumstances of the project at hand.

Sophisticated Science

On every issue, we can apply the scientific method. Researchers can conduct rigorous studies, measure human attributes, and build mathematical models. As the field evolves, an increasing number of analytic tools that have practical application are becoming available.

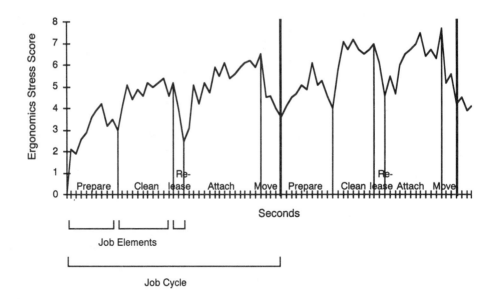

As an example of sophisticated science, ergonomists have developed a variety of computer-based analytic tools for the study of human activity. The graph illustrated here represents output from the ERGAN™ system.

Common Sense

There is another level of ergonomics, the one that I personally enjoy more. This is the level of common sense, where everyone can make a contribution. As mentioned previously, much of ergonomics can be intuitive and ordinary people can provide many ideas about improving the usefulness of products, particularly if they receive training in the basic principles.

❙ We all modify our surroundings to suit us.

Much depends on how the field is presented, however. If ergonomics is taught as a complicated field of study where a Ph.D. is required, quite naturally people will feel intimidated and that the issues are beyond them. On the other hand, if ergonomics is presented in a more straightforward fashion, then almost everyone can grasp and start applying the information.

Teaching ergonomics can be like any other type of good design. We know that the operation of some equipment can be designed so well that it appears simple to

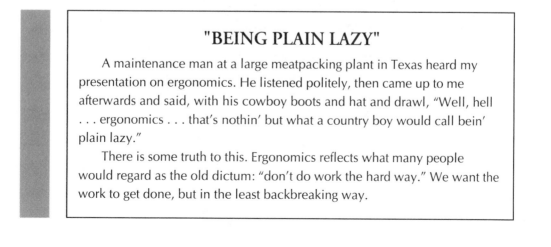

"BEING PLAIN LAZY"

A maintenance man at a large meatpacking plant in Texas heard my presentation on ergonomics. He listened politely, then came up to me afterwards and said, with his cowboy boots and hat and drawl, "Well, hell . . . ergonomics . . . that's nothin' but what a country boy would call bein' plain lazy."

There is some truth to this. Ergonomics reflects what many people would regard as the old dictum: "don't do work the hard way." We want the work to get done, but in the least backbreaking way.

use. Similarly, ergonomics can be taught so well that the process of application appears simple.

Especially in the early stages of an ergonomics project, it is useful to take the commonsense approach. In later stages, after the easy-to-solve issues have been addressed, more rigorous methods of conducting studies and measurements may be useful.

It Does Not Have to Be Expensive

The expense of applying ergonomics can fall within a wide range. At one extreme are expensive, capital-intensive investments. At the other end are changes that can be made with little or no investment.

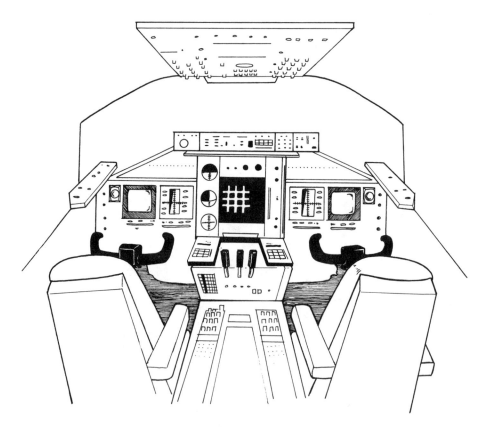

Improving the design of aircraft cockpits may not be cheap, but the paybacks are obvious.

Expensive Long-Term Investments

Ergonomic improvements can sometimes be expensive. New capital-intensive equipment may be required and new facilities built. Elaborate studies may be needed. These improvements may still be cost-beneficial, but may demand large investment.

Low Cost

Improvements need not be expensive. It is often surprising what can be done for low or no cost by using a little imagination (Table 3.1).

TABLE 3.1   *Fifty Ergonomic Improvements Costing under $200*

	Commercially Available	Equipment Upgrade
Office	Copyholder Telephone shoulder rest Telephone headset Adjustable/tiltable computer stand Lumbar support pillow Footrest Wristrest Overhead light diffusers Antiglare screen Printer noise enclosure	Chair adjustability Improved casters Containers with handles Wrist straps Pistol grip handle Power tools Dedicated tools Floor stands
Manufac-turing	Tool balancer Arm balancer Padded edging for workbench Power assists Antifatigue mat Nonskid floor covering Cushioned insoles Material handling carts Lean stand Back support Clamps, vices, other fixtures Rollers, supports, and skids Second handle for tools Antitorque bars Customized tool grips Vibration-dampening tool wrap Task lighting	Store heavy items at waist height, lighter items at higher/lower heights Layout changes Reduce workbench dimen-sions Lazy Susan Align work heights Tilted surfaces Alternate sitting and standing Guide bars, funnels, etc. Improved grips on loads Several size gloves Methods improvement Good housekeeping Good maintenance Floor repairs Understandable signs and in-structions

As an example of a quick fix, this computer monitor has been raised by placing it on top of a phone book. Difficult heights can often be improved by such simple measures.

It Works Well in a Team Process

In recent efforts to apply ergonomics in the workplace, two general approaches have evolved for the process of conducting ergonomic task analysis and problem-solving. These approaches can be referred to as the *individual approach* and the *team approach.*

Individual Approach

The individual approach can be described as a more traditional approach to problem identification and problem-solving. Typically, it involves an outside expert, an in-house engineer, or a top-down-style manager. Usually the activities are performed by a single individual, but sometimes a group of individuals can perform in the same fashion. The key is that the process of identifying and solving problems is done in isolation from end-users, typically production employees and first-line supervisors. The work is done by experts who solve problems for others.

There are a number of advantages to this approach. The process may be easier to implement, since there is no need to train others in principles of ergonomics or in techniques of risk evaluation and problem-solving. Furthermore, there is typically no need for potentially time-consuming meetings or elaborate communications efforts. This approach seems to be particularly useful if precise information on ergonomic problems is sought, that is, a single investigator can take measurements of known issues and describe the problem in detailed, rigorous fashion.

There are also potential disadvantages. The investigator(s) may lack crucial information known by others, particularly by employees about subtle problem areas and by supervisors on constraints to potential improvements. Furthermore, there is often a lack of "ownership" of ideas for improvement by others who are necessary to implement these ideas, whether these other personnel are production employees, or in the case of an outside expert doing the investigation, in-house engineers and managers. Thus, the recommendations for improvement may not be feasible or simply not accepted by workplace personnel.

Team Approach

This approach involves establishing a team that includes the investigator. The composition of this team can vary widely with each workplace. Sometimes a "team" can simply be a few people talking informally at the workstation. Other times, the team is more formal, including a cross section of people:

- management, including top and first line
- production employees and union representatives
- maintenance and engineering
- health care providers
- equipment vendors
- outside experts, such as professional ergonomists

In particular, as specific tasks are reviewed and improvements sought, there is increased participation of employees.

The advantages to this approach are several. Often a fuller picture is gained of the issues because of input from people with various perspectives. Consequently, there can be a higher quality of understanding of the problem and potential improvements. Furthermore, as a result of participation, there is often an increased

YOU NEED TO TALK TO USERS

Tool and workstation design often differs from other aspects of facility planning in an important way. It is quite possible, for example, to have an electrician review the needs of a workplace and design a wiring system that performs perfectly adequately without talking to any users. Setting up a workstation is different; there are too many nuances in the design that are only known to people who have actually done the work. For the best designs, communication between users and designers is needed.

acceptance of the proposed changes, making it easier to implement these changes. Finally, because of the early involvement of people, it is often easier to refine the changes and make follow-up workstation modifications.

There are also disadvantages to this approach. More time for meetings and coordination is typically required. The process, because it involves more people, can be more difficult to manage. Finally, because it has not been a traditional process, some effort is required to encourage people to go through the steps.

Both of these approaches—individual and team—can work, although on occasion, circumstances in individual workplaces may dictate that only one or the other

SUSAN'S DESK

Susan, a secretary, learned that she would soon have a new boss. The following Monday, Susan arrived at work and found that her desk had been reorganized.

"Who did this?" she asked, astonished.

"I did," said the new boss, coming out of her office over to Susan's desk.

"Why?" Susan asked, yet more dumbfounded.

"Because it's better that way," replied her boss.

"I don't care if it's better!" was Susan's reaction. "That's my desk! I'm changing it back."

Susan's reaction is readily understandable. It is unusual in an office setting for this type of change to occur without the employee's involvement.

Yet, in the manufacturing environment, it is common to change things without employee involvement or even knowledge. It often happens that employees show up on a Monday morning to find their workstations rearranged. They have not had opportunity to provide their own ideas, let alone be given the common courtesy of being informed prior to the changes.

Fortunately, times are changing and employees are involved much more than previously. This anecdote, however, highlights the importance of involving people and using a team concept, especially when changing a workstation.

approach be used. However, as a general rule, the team approach provides the best results. Many managers expect an individual expert approach from an ergonomics consultant, but once they experience the team process in action, they typically appreciate its merits.

There Are Many Ways to Solve Problems

The phrase "options for improvement" is a good one to keep in mind when trying to address ergonomics issues.

Multiple Options

People often presume that there is one—and only one—solution to a problem. Experience, however, shows that there are almost always multiple options for making improvements.

For example, to raise the height of a computer monitor (the computer screen) you have a variety of options:

- place the monitor on a phone book
- build a wooden monitor stand
- purchase an adjustable monitor stand
- purchase an adjustable monitor arm (wall-mounted)
- purchase an adjustable monitor arm (desk-mounted)
- place your desk on blocks (or casters) to raise it up
- purchase an adjustable-height desk
- purchase a special computer workstation that has separate adjustments for the monitor, the keyboard, and hard copy (and there are many options available on the market from different vendors)
- lower your chair

Continuous Improvement

Furthermore people often presume that once this ergonomic feature is implemented, the issue is fixed for all time. The phrases you hear are "this is ergonomically correct" or "that is not ergonomic."

In reality, *ergonomic* is a relative term. For the most part, the word *improvement* is a more suitable term than *solution,* since once a task is made better, it can almost always be improved on at some point.

Thus, "ergonomically correct" depends to a large degree on what you compare it with:

- A metal folding chair is more ergonomic than sitting on the ground.
- An old-style padded office chair is more ergonomic than a folding chair.
- A new "ergonomic" office chair (pneumatic adjustment, five-star base, and other features) is more ergonomic than the old-style padded chair.

All of these can work, depending on the circumstances. There is no one automatically best choice.

Why These Distinctions Can Save You Money

Many standard recommendations for "ergonomically correct solutions" can require unnecessary expense. There may be other equally valid improvements possible that are both less expensive and more effective.

The problem is one of not taking the time to brainstorm all of the options. Usually, people think of one or two ways to make an improvement and stop there, not considering any of the score of other possibilities.

Furthermore, for cost reasons, sometimes it is appropriate to take the incremental approach: to make an affordable improvement now, then wait to make an even more substantial improvement later. A good example would be to provide a small floor stand to enable a shorter person to reach high overhead. Then, the next time the work area is to be renovated and new equipment installed, redesign the layouts to eliminate the overhead reach altogether. Often, the costs of these types of changes are negligible when new equipment is installed. (Bear in mind that, depending on the situation, it might be best and most cost-effective to eliminate the overhead reach immediately. The point is to explore all of the options first, a step too often neglected.)

Pitfalls in Purchasing Ergonomic Products

Warning

- "Ergonomic" products that are used inappropriately are not ergonomic.
- Any product that leads to improvements can be "ergonomic." Be creative. Ideas for improvement can come from anywhere.
- There is no certification or testing procedure to determine if a product is truly ergonomic—buyer beware!

To be able to evaluate products, you must understand the underlying principles of ergonomics.

Be Creative

There can be countless options for making improvements. For example, before purchasing a lift table to reduce the risk of injury from lifting, determine if there are radically different ways of accomplishing the same task. You might just be able to identify a completely new and better way of doing the job.

Improvements can be found through:

- traditional in-house engineering improvements
- homemade devices

- creative thinking
- changing the work process or entire work area

MORE ON "WHAT IS ERGONOMICS?"

People versus Machines

Many ergonomists have found it useful to contrast the strengths and weaknesses of both people and machines. Although sometimes these statements seem self-evident, at other times they help put issues in perspective and give guidance to strategies and decision-making.

People . . .	*But Machines . . .*
Are at their best when they do a wide range of variable motions and tasks.	Have difficulty doing complex motions and varied tasks.
Are creative, can plan and invent.	Need programmed, preset responses.
Can react to unpredictable events, identify and choose options.	Perform specific functions within narrowly defined limits.
Are better at interpreting data.	Are better at processing high volumes of data.
Machines Can Easily . . .	*But People . . .*
Do highly repetitive motions at high force and speed.	Are limited, wear out, and lose accuracy quickly.
Detect small variations in routine tasks.	Are easily distracted and make errors.
Perform accurately in endless repeated tasks.	Are inconsistent, and need change in stimuli.
Work in hostile environments (e.g., toxic, radioactive, extreme temperatures).	Are highly affected by the environment.
Have limitless design.	**For better or worse, come with a fixed physical design.**

The Scope of Ergonomics

In the same manner that a chemist can view all of the world in terms of chemistry, the ergonomist can view virtually all human interaction as ergonomics. Anything that helps humans expand on their capabilities or overcome their limitations can be viewed as components of this field of study.

The following schematic summarizes the scope of ergonomics:

Motivating Factors

Improve human well-being

safety

comfort

Improve human performance

productivity

quality

Reduce costs

injuries

errors

Meet human resource trends

aging

slower-growing labor pool

increasing employee expectations

Meet regulations

OSHA

ADA

VDT

Contributing Disciplines

engineering

psychology

medicine

physiology

anatomy

anthropology

industrial design

Principles of Ergonomics and Design Guidelines

physical

cognitive

▼

The Process of Ergonomics

task evaluations
prioritization
problem-solving
evaluation
continuous improvement

▼

Applications

tools, furniture, workstations, and offices
production processes and workflow
displays and controls
instructions
labels
communications
decision-making
home appliances and consumer products
transportation systems
computers
sports and leisure activities

▼

Results

improved human well-being
increased efficiency in doing tasks
reduced injuries
fewer errors and accidents
lower costs
innovation
increased sales
improved profits

TERMINOLOGY: OXYMORONS

Human Ergonomics

Some people, primarily health care providers, have begun to use the term *human ergonomics* to refer to exercise and fitness programs that keep people in shape to better withstand the rigors of daily living. This term should be discouraged for a variety of reasons.

- In one sense, the term *human ergonomics* is redundant. Ergonomics by definition involves humans; if there is no human involved it cannot be ergonomics.
- In the sense that these health care providers intend it, the term is an oxymoron. Since the point of the fitness programs is to change people rather than focus on the design of tools and equipment, the term *human ergonomics* is really the antithesis of ergonomics.

This is not to say that fitness is not a good thing—clearly it is. Nutrition and getting a good night's sleep are also important for health and well-being, but no one would ever call these practices ergonomics. Exercise is good, but it is not ergonomics.

Ergonomic Hazards

This term has also begun to crop up, but it is another oxymoron that should be discouraged. *Ergonomic hazard* has been used to refer to risk factors for cumulative trauma. By definition, ergonomics refers to design concepts for improving work—solutions, not hazards.

What Isn't Ergonomics

It is sometimes easier to explain what ergonomics is not, in order to present the scope of the field. The following statements provide some boundaries on what ergonomics is not:

- If a human is not part of the system or interaction, it is not ergonomics.
- If there is no underlying goal to make improvements or affect design, it is not ergonomics.
- If the concern is for the human alone, without regard to tools and systems, it is not ergonomics.

TABLE 3.2   *What Is and What Isn't Ergonomics*

Application	What Is Ergonomics	What Isn't Ergonomics
Automobiles	Easy access in and out of car door; comfortable seating; understandable dashboard; standardized and stereotyped location of controls; easy acess to engine to make repairs; air-conditioning	Color and evenness of paint finish; fuel efficiency; engine type and horsepower
Fishing	Comfortable seating (bass boat seat or soft soil at fishing hole); well-organized tackle box; easy access to landing net; sharp fillet knife	Skill in catching fish; enjoyment with fishing companions; pleasure from eating fresh fish
Hotel	Understandable directions to rooms; efficient check-in and check-out systems; standardized and stereotyped shower controls	Attitude of service staff; cable TV; wall decorations
Playing guitar	Wrist posture while compressing strings; noise exposure level	Happiness and self-fulfillment from playing music
Air traffic control	Communications between pilots and air traffic controllers during takeoff and landing	Social relationships between pilots and air traffic controllers after work
Highway intersections	Clear signs and direction; standardized and obvious traffic patterns	Height of overpasses; durability of road surface; land use issues
Machines, in general	Using a machine in a new way, or expanding the capabilities of a machine, to reduce human labor	Improving the purely internal mechanics of a machine (such as making a machine more energy efficient, where there is no direct human–machine interaction)
Personal self-improvement	Training in work methods to perform tasks more efficiently	Improving self-esteem or overcoming emotional problems (where there is no involvement with a tool or task)
Exercise	Providing exercise breaks as part of the work routine to relieve monotony or static postures; exercise equipment designed to help you exercise more efficiently	Staying in shape

(cont.)

TABLE 3.2   *(continued)*

Application	What Is Ergonomics	What Isn't Ergonomics
Work organization	Assembly lines, work cells	Company picnics
Training	Techniques of training: mnemonics, redundancy, immediate practice in applying concepts	Content of the training

The variety of applications given in Table 3.2 illustrate what is and what is not ergonomics. The point here is not to draw fine lines; the distinctions between many fields of study are blurred and often a matter of perspective. The examples in Table 3.2 simply provide a useful explanation of ergonomics and clarify educational points, albeit somewhat tongue in cheek.

Note that all of the items in the right-hand column of Table 3.2 are clearly important. Furthermore, many items have to do with design criteria and some have to do with humans interacting in a system. Yet, none can be designated as part of ergonomics because in one way or another they: (a) do not involve a human, (b) do not involve a tool, or (c) do not relate to design or improving the efficiency of humans in a system.

"Old-Fashioned" Ergonomics

As mentioned in Chapter Two, in some ways ergonomics is nothing new. From the very first tool of the Stone Age, humans have tried to find better ways of working, taking advantage of our talents and using machines to overcome our limitations. What is new, however, is a more scientific and systematic approach to understanding human anatomy and physiology, and learning about human differences, limitations, and reactions.

The above list is somewhat facetious, but the point is that ergonomics is not necessarily anything esoteric or extravagant. On the contrary, ergonomics addresses the core of human economic and technological development.

"Modern" Ergonomics

During World War II, scientists and engineers began to consciously study human capabilities and limitations with the goal of improving the design of military equipment. For example, the speed and maneuverability of fighter aircraft

GREAT ERGONOMIC IMPROVEMENTS IN HISTORY

Improvement	Ergonomic Benefit
• stone ax	• reduced exertion; overcame human limitations
• wheel	• reduced exertion
• language	• more precise communications
• agriculture	• more efficient system of organization
• sundial face	• improved clarity in presenting information
• road sign	• improved clarity in presenting information
• chair	• improved comfort; reduced static load
• two-handed scythe	• improved posture, reduced exertion; extended human capabilities
• city grid system	• improved efficiency in locating and transporting
• power sources:	• reduced exertion and repetitive tasks; overcame human limitations
• waterwheel	
• steam	
• electricity	
• computer	• reduced repetitive tasks (e.g., mathematical calculations); overcame human limitations (e.g., searching large data bases)

began to overwhelm the ability of pilots to keep up with fast-moving events. In response, aircraft designers began to involve psychologists in an effort to rethink the design of the cockpit, to make it what we would now call more user-friendly.

Another example relates to improving the accuracy of antiaircraft weaponry, namely, the ability of human operators to track aircraft and direct fire. Studying

Nineteenth century ergonomics. Using a one-handed sickle was literally a back-breaking task, a source of misery for generations of peasants. In the nineteenth century the two-handed scythe—a great ergonomic device—solved many problems. Note the features: (1) upright posture for lower back, (2) "neutral" posture for wrists, (3) adjustable hand grips, and (4) better use of large muscle groups. The two-handed scythe was an unconventional design (at the time), but the result was greater efficiency: more work done with less effort and better control over the tool.

these skills and decision-making patterns for this purpose continued in the following decades and aided in the development of antiaircraft missile systems, smart bombs, and other advanced weaponry.

These steps represented the birth of modern ergonomics. Ever since, ergonomics has played a strong role in the aerospace industry. Indeed, one of the most complete data bases on human dimensions was compiled by NASA for use in designing space capsules.

In subsequent decades, ergonomics has expanded into a variety of other applications. The field has gone far beyond following basic human nature and has emerged as a science.

CUSTOM TOOLS—A TRADITION

Sometimes when I talk about customizing hand tools to fit individual hands or developing dedicated tools to be used just for certain tasks, I am met with skepticism. The reaction is as though I am advocating an extremely radical idea or suggesting pampering a generation raised in the easy life.

That is not the case, however. I offer as an example my grandfather, born in a log cabin in the woods of northern Minnesota to Swedish immigrant homesteaders. His family was among the first pioneers in that area and certainly accustomed to hard work and self-reliance.

His toolshed provides evidence of an old tradition of designing for the individual user. Virtually every tool was custom-built to his own personal dimensions. For much of his life, when he wanted, say, a new hammer, he would go to town and buy just a steel hammerhead. That was the only part he could not make himself. The handle he could make. And that he did—for his own hand.

Moreover, the walls were filled with gadgets and devices that he had concocted to do one chore or another. Many looked like they had been used only a few times. But he took the time to construct the right tool for the right job.

It seems it is only recently, with the advent of mass-produced tools, that we have gotten away from customizing tools to individuals and specific tasks.

The point I want to make is that much of ergonomics is a part of our heritage. The only difference is that now we can study and measure things in a way that we could not before. We are applying the basic principles to new areas, such as communications and decision-making, and we are taking actions that were somewhat subconscious in the past and turning them into conscious design techniques.

The Future of Ergonomics

All in all, ergonomics is still a developing field. Often, the only guidance that ergonomics can provide is a set of general principles, or a series of rules, with many details still unknown. The future, however, should bring much more sophistication to the evaluation and design process:

- improved ways to characterize work and human performance
- better measuring instruments and techniques
- stronger data bases from which to draw

HOW "MODERN" ERGONOMICS DIFFERS FROM "OLD-FASHIONED" ERGONOMICS

- A systematic approach is used, rather than relying on natural, but haphazard impulses. The scientific method is applied.
- Modern ergonomics provides an improved mind-set to deliberately and consciously apply principles, rather than follow intuition.
- Many analytic techniques and tools have been devised by modern-day ergonomists to permit a process of design.
- Sophisticated measuring instruments have been and are being devised.
- A broad data base of knowledge has begun to amass.
- More rigorous design criteria are being developed.

EXAMPLES OF APPLICATIONS

The following examples provide an introduction to the widespread applications of ergonomics. One of the purposes for providing all these examples is to indicate that ergonomics is not an esoteric field, but rather something that affects all our lives.

Aviation

In recent decades, a great deal of ergonomics research and development has taken place in the aerospace industry, as well as in the military in general. Pilots have been the beneficiaries of the most concerted ergonomics efforts ever, regarding the design of the cockpit. Here, ergonomics has addressed both the physical fit of the pilots into a small space and the ability to grasp complex information quickly. Decision-making must be made simple since the pilot needs to gather information, make correct decisions, and perform complex interactions, often in a matter of seconds. Nowhere is this more critical than with military pilots in fighter aircraft.

Computers

Likewise, significant work in ergonomics has been accomplished in the fields of computer equipment and software design. In particular, the human–computer interface has been the subject of much attention. Of crucial importance is determining the most effective way for the computer to present information to us—screens, windows, icons, and so forth. Similarly, we need to know what is the best system for controlling the computer—keyboards, mice, touch screens, or something else. Ergonomics has addressed all of these issues.

Automotive

Automobile design is receiving increased attention from an ergonomic point of view. For years, cars have had adjustable seats and mirrors to take into account the wide variation in the size of drivers. Presently the design of seating to provide better lower back support is occurring. Layouts of the dashboard and locations of control buttons have been improved in some cases (and unfortunately worsened in others).

Consumer Products

Other consumer products ranging from toothbrushes to cameras have benefited from improved attention to ergonomics. Even the difficulty in programming a VCR—what had become a cliché in our society—has begun to be improved.

Manufacturing

In recent years the manufacturing industry has been the focus of considerable ergonomics effort. An area of traditional investigation has been reducing accidents and errors through better layout and arrangements of control panels and through better design of warning signals. More recent are efforts to prevent cumulative trauma, primarily by applying ergonomic principles to reduce exertion, improve working postures, and reduce repetitive motions.

Guidelines to determine when workloads are excessive, both physically and mentally, have also been sought. Finally, ergonomics is beginning to reach out to new areas such as inspection tasks; humans often perform poorly when attempting to detect minor variations in routine tasks and there is a need to find alternatives to present practices.

Maintainability

An important application of ergonomics is designing equipment so as to be easy to maintain. Some issues are purely physical, such as having clear access to reach needed areas. Other issues are of the cognitive type.

A good example is the improvements that have been made in recent years for simple repair of photocopying machines, such as removing jammed paper. As the machine's front cover is opened, clear direction is given regarding ways to uncouple the equipment, remove the jammed paper, and put things back together again.

What was once incomprehensible to all but trained technicians is now doable by ordinary office personnel.

Service

Ergonomic issues in service industries are also being addressed. A primary issue here is eliminating confusion for customers, for example, how to fill out a form, which line or window to go to, how to exit a parking ramp. These are the kinds of frustrations we often experience in our everyday lives. Ergonomists are developing analytic techniques that provide excellent guidance to make directions clear and reduce the confusion.

Office

As a final example, the office environment has been the subject of much recent ergonomics effort. Issues have centered primarily on adjusting the office to the use of computers, but range from design of chairs and lighting to overall work organization.

Computers: More Examples of the Application of Ergonomics

A more detailed look at computers can shed further light on the concepts involving ergonomics. The computer itself is an ergonomic tool, since it (a) takes advantage of and expands the capabilities of humans and (b) overcomes human limitations in performing high volumes of rote mental tasks. However, a number of ergonomic issues have arisen in the design of computers and, in fact, created some new problems for computer users. Fortunately, these issues have been resolved (or can be) through the application of ergonomic principles.

Computer Screens

Early computers that used punch cards for data input and output were cumbersome for human use. Connecting a monitor to the computer vastly simplified this interaction between humans and the machine. The advantages of computer screens are in retrospect so obvious that it is difficult to even remember that the first computers did not have screens.

Subsequently, a variety of issues involving these screens themselves had to be addressed. Made with reflective glass, the screens introduced a highly polished surface into the workplace. Given that the operator's job involves looking directly at this reflective surface, lighting and glare began to affect large numbers of people in a way that had not occurred before.

Fortunately, manufacturers learned to process the screen surface and to use materials in ways that reduce glare. Devices that can be positioned over the screens to further reduce the problem also became available.

Furthermore, computer users and office planners are learning how to lay out work areas to keep the screens away from light sources. We are learning to change the nature of lighting to keep light levels lower, to use indirect lighting, and to use smaller lights—"task lighting"—that illuminate only the specific objects such as paper copy that we need to see clearly. Incidentally, these are the same techniques that new owners of television sets in the 1950s learned to use. At that time, magazine articles listed steps one could follow to overcome the lighting issues concerning TV sets: keep the TV in a separate dimly lit room, use background lighting, and draw window curtains. This is much the same as we see today for computer use.

Software

Early computer programs required memorization of complex commands that were hard to remember and often inconsistent. Today the best programs are almost instinctive to operate, especially if users have had some previous familiarity with computers. Users are led in often self-evident ways into doing the right thing to accomplish what they want. The results of good ergonomics in writing software have become clear: reduced learning time, increased performance speed, lower error rates, and improved user satisfaction.

ERGONOMICS IN SOFTWARE DEVELOPMENT

[Ergonomics] was once seen as the paint put on at the end of a [software] project, but is now understood to be the steel frame on which the structure is built.
Schneiderman (1987)

It is now quite evident that computer users, once they become used to easy systems, quickly reject unfriendly programs and move on to other products. The implications are clear for businesses that produce software, as they should be for producers of any product in today's marketplace.

Keyboard

As with computer monitors, the keyboard was a great ergonomic step forward in increasing the friendliness of computers. Attaching the keyboard linked humans directly to the actions of the computer. Furthermore, the keyboard was an "input device" with which people were well familiar. As with monitors, it is difficult now to envision how early computer technicians performed their tasks without keyboards.

Unfortunately, problems also arose with use of this device. For typists, the keyboard in conjunction with a computer word-processing program permitted much faster typing speeds than did a typewriter. Additionally, many of the original keyboards were quite thick and inadvertently placed the wrists in poor postures. Thus, cumulative trauma disorders of the wrist began to appear in a way that was not seen earlier.

The layout of the keys on the keyboard has even become an issue. When the typewriter was invented in the late 1800s, there was a mechanical bottleneck in the machines exacerbated by the need for speed. Based on subsequent studies, the keys were oriented to separate routinely-used combinations. The resulting layout was slower and a bit more awkward for typists, but kept jams from occurring. The key layout is known as the QWERTY system, after the letters of the keys on the uppermost row.

We have inherited a system that was designed to solve a problem that no longer exists. With computers, all of the linkages are electronic—mechanical jamming is

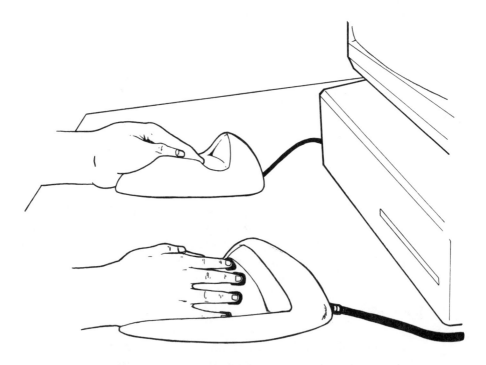

New input devices are being designed with ergonomics in mind.

precluded. Furthermore, there is no particular need for a *board* arrangement of keys anymore. With electronics, the keys can be mounted on devices of any shape.

Currently, a great deal of experimentation is taking place. Alternative keyboard layouts have been designed. One system is the Dvorak keyboard, named after its inventor, which has been shown to be easier to learn and faster to use, yet reduces finger motions. Additionally, unconventional keyboards have been developed, such as hemispheric and A-frame devices. An even more radical departure is a pair of mittenlike devices onto which the hands are placed, with keylike sensors that require almost no movement to activate. All of these unconventional keyboards have been designed either to place the wrist in a less injury-prone posture or to reduce finger exertion and repetition. In some cases these keyboards also provide convenient locators for the hands to prevent inadvertent typing on the wrong sets of keys.

Finally, even the traditional devices have been greatly improved since the initial designs. Keyboards have become thinner and thus place the wrist in better posture and new devices such as wrist supports have been introduced. Design of the individual keys has also been improved via a slimmer design, requiring less finger activity.

Other input devices, such as the mouse, graphic tablets, and voice recognition, have also proven effective and made computers easier to use. Many design issues regarding these devices—primarily the resulting posture of the wrist—have been addressed through ergonomics, or in some cases have yet to be addressed.

Office Furniture

Similarly, computer technology has had an impact on furniture design in two important ways:

- First, the computer is immobile and people generally have lost the opportunity for flexibility that is often found in traditional work. Instead of employees being able to shift items such as books and paper tablets around to suit themselves, they must now adjust their bodies to fit the computers. Reaches, postures, and comfortable visual distances are affected, mostly to the detriment of employees and the quality of their work.
- Second, because people can do so much more work and so many more kinds of activities on computers, we simply end up staying in the same posture for longer periods of time. Thus, good furniture becomes more important than ever before. In the past we often put up with inadequately padded seats and nonadjustable chairs because we did not need to stay in one place or in one position for long periods of time. Now, because of the computer, we have to pay attention to these issues.

Fortunately, the market has again responded with innovation and new products. Ergonomic chairs and workstations are now available, as are new devices such as adjustable arms for computers and keyboards. Furthermore, computer technology is evolving rapidly. For example, the types of screens that are used for laptops are thinner and lighter, yet of excellent image quality. The computers themselves are simultaneously becoming more powerful, yet weigh less and require less space. Computers in the near future may be mounted on highly flexible arms or in other ways made so portable that we will be able to instantaneously adjust the computer, keyboard, and screen to fit us exactly (some of this book was written using a laptop computer while reclining in a hammock).

Customer Interactions

As a final note, use of computers has interfered with many normal face-to-face human interactions, which can adversely affect service and business. In hotels and banks, for example, the location of the computer screen forces the employee to look away from the customer. The employee becomes more involved with the computer than with the human. Although computers are a necessary tool, better layout of counters and other ergonomic applications could reintroduce the human element.

section 2 Management Issues and Strategies

chapter FOUR

UNDERSTANDING CUMULATIVE TRAUMA

One of the most perplexing occupational safety and health issues confronting managers today is the seemingly sudden appearance of a class of injuries known as cumulative trauma disorders (CTDs). This chapter defines these disorders and explains why the rising incidence of CTDs ought not be so mystifying. The costs of cumulative trauma are discussed and guidance offered for the prevention of CTDs.

The number of recorded cases of CTDs is rising in the United States, as shown in the figure below. These national data verify what many managers have experienced in their own workplaces. From all indications, the trend is similar in other advanced industrial nations. This sudden increase is the reason why the Occupational Safety and Health Administration (OSHA) is targeting CTDs and why there has been such controversy on this issue.

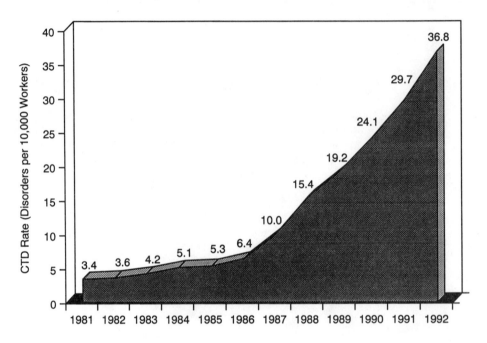

The rate of work-related upper-limb CTDs in general industry in the United States is rising rapidly (*Source:* Bureau of Labor Statistics, 1993.)

The sudden increase in reported CTDs has caught many managers off guard. The common experience in many, if not most, workplaces is that each operated successfully for decades without any recorded disorder until the present epidemic began suddenly in the 1980s.

Why has this occurred? Many managers have been baffled by this question. Some people have suspected widespread malingering or have viewed the trend as evidence of a weakening society. However, there are clear reasons for these increasing rates and current research reveals many of the answers.

WHAT IS CUMULATIVE TRAUMA?

In lay terms, CTDs can be described as wear and tear on joints and surrounding tissue because of overuse. Every joint in the body can potentially be affected, but the lower back and upper limbs are the areas that receive the most injuries. Cumulative trauma can be contrasted with "acute" trauma, that is, instantaneous injuries such as those caused by cuts or falls. Cumulative disorders *accumulate* through time, hence the term.

CTDs are quite widespread in the general population. Most of us have experienced a CTD at one point or another in our lives in at least a mild form, such as low back pain.

Many CTDs are sports- or recreation-related. Baseball pitchers, for example, develop various disorders in their pitching arms from the cumulative effect of forceful, repetitive motions done with the arm in an awkward position. In fact, these problems limit the number of pitches per game that can be thrown and often end lucrative careers. Baseball catchers also suffer glove-hand problems, resulting from

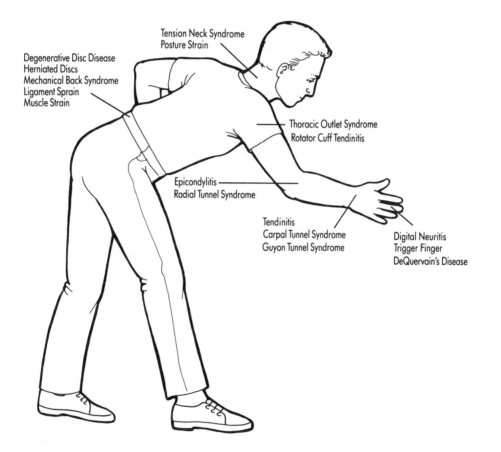

Tension Neck Syndrome
Posture Strain

Degenerative Disc Disease
Herniated Discs
Mechanical Back Syndrome
Ligament Sprain
Muscle Strain

Thoracic Outlet Syndrome
Rotator Cuff Tendinitis

Epicondylitis
Radial Tunnel Syndrome

Tendinitis
Carpal Tunnel Syndrome
Guyon Tunnel Syndrome

Digital Neuritis
Trigger Finger
DeQuervain's Disease

Some of the wide variety of cumulative disorders. Medical terminology varies by the particular part of the body affected and the precise tissue damaged, but from a lay point of view, most of these fall into a few basic categories, such as tendinitis or nerve disorders.

the repetitive and highly forceful impact of catching innumerable pitches. Long-distance runners suffer a host of lower-limb CTDs. The origin of terms like *tennis elbow* and *golfer's elbow* is obvious.

Similarly, in the workplace many employees can experience a CTD in one form or another during their working lives. Virtually all occupations can be affected, although some clearly more than others.

Most often, cumulative disorders are mild and temporary. But in their more severe forms, CTDs can be very painful and sometimes permanently disabling. The disorders can interfere with all aspects of daily life, sometimes to the point where the simplest of work tasks, household chores, and social events can become unmanageable.

Although there are differences based on the exact type of disorder and the part of the body affected, the symptoms of CTDs in general include:

- soreness, pain, and discomfort
- "burning" sensations
- limited range of motion
- redness and swelling
- stiffness in joints
- weakness and clumsiness
- numbing, tingling sensations ("pins and needles")
- popping and cracking noises in the joints

GLOSSARY OF CUMULATIVE TRAUMA DISORDERS

Hand and Wrist

- **Tendinitis**—Inflammation of a tendon.

- **Synovitis**—Inflammation of a tendon sheath.

- **Trigger finger**—Tendinitis of the finger, typically locking the tendon in its sheath causing a snapping, jerking movement.

Neck and Back

- **Tension neck syndrome**—Neck soreness, mostly related to static loading or tenseness of neck muscles.

- **Posture strain**—Chronic stretching or overuse of neck muscles or related soft tissue.

- **Degenerative disc disease**—Chronic degeneration, narrowing, and hardening of a spinal disc, typically with cracking of the disc surface.

Hand and Wrist

- **DeQuervain's disease**—Tendinitis of the thumb, typically affecting the base of the thumb.

- **Ganglion cyst**—Synovitis of tendons of the back of the hand causing a bump under the skin.

- **Digital neuritis**—Inflammation of the nerves in the fingers caused by repeated contact or continuous pressure.

- **Carpal tunnel syndrome**—Compression of the median nerve as it passes through the carpal tunnel.

Elbow and Shoulder

- **Epicondylitis** ("tennis elbow")—Tendinitis of the elbow.

- **Bursitis**—Inflammation of the bursa (small pockets of fluid in the shoulder and elbow that help the tendons glide).

- **Rotator cuff tendinitis**—Tendinitis in the shoulder.

- **Radial tunnel syndrome**—Compression of the radial nerve in the forearm.

- **Thoracic outlet syndrome**—Compression of nerves and blood vessels under the collar bone.

Neck and Back

- **Herniated disc**—Rupturing or bulging out of a spinal disc.

- **Mechanical back syndrome**—Degeneration of the spinal facet joints (part of the vertebrae).

- **Ligament sprain**—Tearing or stretching of a ligament (the fibrous connective tissue that helps support bones).

- **Muscle strain**—Overstretching or overuse of a muscle.

Legs

- **Subpatellar bursitis** ("housemaid's knee" or "clergyman's knee")—Inflammation of patellar bursa.

- **Patellar synovitis** ("water on the knee")—Inflammation of the synovial tissues deep in the knee joint.

- **Phlebitis**—Varicose veins and related blood vessel disorders (from constant standing).

- **Shin splints**—Microtears and inflammation of muscle away from the shin bone.

- **Plantar fascitis**—Inflammation of fascia (thick connective tissue) in the arch of the foot.

- **Trochanteric bursitis**—Inflammation of the bursa at the hip (from constant standing or bearing heavy weights).

Most CTDs of the hand and arm fall into two main categories: tendinitis and nerve compression.

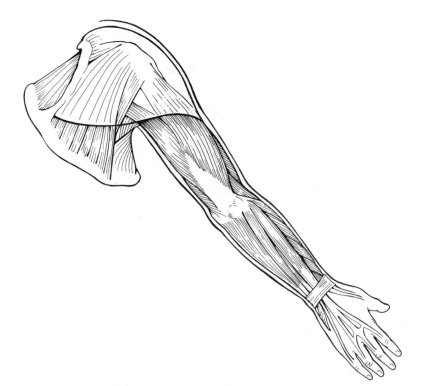

| The shoulder, arm, and hand (the "upper extremity") can be affected by CTDs.

Tendinitis

Tendons are found throughout the body, serving as links that connect muscle to bone. The tendons come into play every time a muscle is used to perform a motion with a bone structure. In many areas, these tendons slide through sheaths, much like a cable in a bicycle brake.

Along the arm, tendons are found in the shoulder, elbow, and throughout the hand and wrist. For example, the tendons in the wrist connect the fingers with the muscles in the forearm. Every time there is finger, hand, or wrist activity, these tendons slide back and forth inside the hand and forearm. The tendon action can be easily seen by looking at the palm side of your wrist while opening and closing your hand.

As with any other moving part, overuse of the tendons can create friction, which in turn can cause (1) wear and tear and (2) expansion or swelling. When the tendons or their sheaths swell, and there is pain and tenderness, we call it "tendinitis."

The overall effect is like pulling a rope over a pulley. If the rope is pulled a few times, not much happens to it. But if the rope is pulled thousands of times per hour, day after day, the rope can begin to wear. The same can happen with tendons.

Nerve Compression

Nerves are found throughout the body. In the case of the hand and arm, three major nerves run from the spinal column through the shoulder and arm to the fingers. At several points it is possible for the nerves to be compressed. Sometimes this pinching can be caused by making certain awkward motions or assuming certain postures. Other times the compression is caused by the swelling of nearby ten-

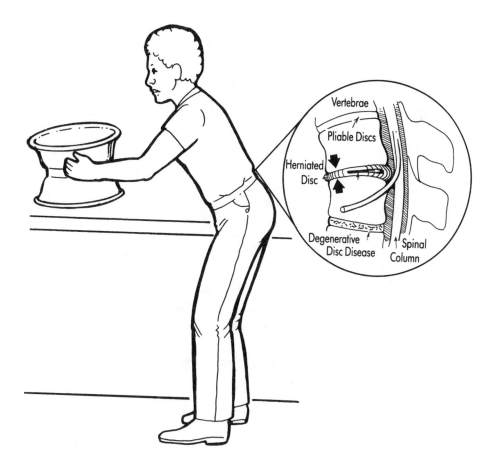

The lower back can be affected by a variety of cumulative disorders.

dons. Another possibility is that factors such as awkward postures and static load reduce blood flow to the nerves, causing damage.

A common example of nerve compression is carpal tunnel syndrome. Inside the wrist is a small channel or tunnel, known as the carpal tunnel (*carpal* is the Latin name for "wrist"). One side of the tunnel is formed by a group of bones on the back of the hand. The other side is formed by a band of ligaments across the base of the palm. Inside this channel runs one of the major nerves and a number of tendons. If this nerve is injured either by swelling of nearby tissue or by reduced flow of blood, it can result in a form of paralysis, with numbness and tingling sensations in the fingers. This is carpal tunnel syndrome, the classical symptom for which is "pins and needles" in the hands at night.

Back Injuries

The back is also vulnerable to cumulative trauma. We often think of a back injury as caused by a single event, like lifting a heavy load. However, more than likely the injury is one result of the cumulative effect of bending, twisting, or excessive sitting or standing. A single event may trigger the injury, or it may be merely "the straw that broke the camel's back."

The spinal column consists primarily of bones (the vertebrae) separated by pliable discs. Through the center of this column run the nerves of the spinal cord.

The discs provide cushioning between the vertebrae—they enable us to bend and twist. Unfortunately, with enough bending and twisting, especially while car-

---

TERMINOLOGY

1. Cumulative trauma is also referred to by a variety of other terms, such as *musculoskeletal disorders, overuse syndrome,* or *repetitive motion disorders.* The terms are not always completely synonymous, in that disorders such as hearing loss, eyestrain, and joint diseases like arthritis are sometimes included in this category. However, in the context of ergonomics, the terms usually refer to the type of overuse injury described here.

2. These disorders are sometimes called *illness.* The origin of this terminology is the format in which they are recorded on the OSHA 200 Log of Injuries and Illnesses. Acute problems are classified as injuries and chronic problems as illnesses. In this framework, CTDs fall into the illness category and consequently are sometimes called *illnesses.* However, this terminology does not fit the common understanding of what an illness is, and ought to be discarded. *Injury* is fine, and *disorder* is best.

rying a load, the discs can suffer wear and tear. After a time, the discs can narrow, harden, and the surfaces fissure and crack, what we call "degenerative disc disease." Once weakened, the discs can bulge out, strain, or herniate (often incorrectly called "slipped disc").

There are a variety of other ways the back can be injured by overuse. Examples are ligament sprain, muscle strain, or a condition known as mechanical back syndrome.

Risk Factors

There are several factors that can increase the risk of developing CTDs. The more factors involved and the greater the exposure to each, the higher is the chance of developing a disorder.

Factors	Risk
Working conditions Physical	
Repetition	The number of motions made per day by a particular part of the body.
Force	The exertion required to make these motions.
Awkward postures	The positions of the body that deviate from neutral in making these motions, primarily bent wrists, elbows away from their normal positions at the side of the body, and a bent or twisted lower back.
Contact stress	Excessive contact between sensitive body tissue and sharp edges or unforgiving surfaces on a tool or piece of equipment.
Vibration	Exposure to vibrating tools or equipment, whether a hand-held power tool or whole-body vibration.
Temperature extremes	Exposure to excessive heat or cold.
Work organization	
Stressful conditions	Certain stressful situations related to management and administrative systems. (This topic is addressed in detail in Chapter Seven, Work Organization.)
Personal issues	
Off-the-job activities	The things that individuals do at home and in leisure activities can also contribute to cumulative trauma.
Physical condition	Poor personal fitness can play a role in the development of some (if not all) kinds of CTDs.
Other diseases	Factors such as gout and diabetes mellitus can also be implicated.

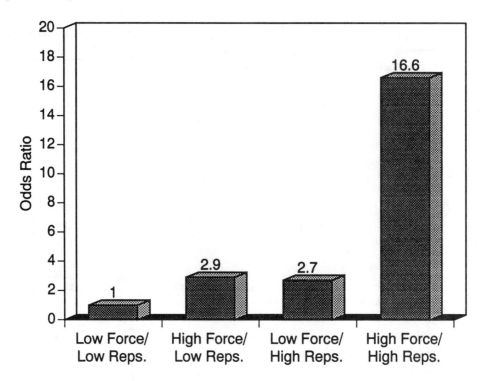

Wrist Disorder Rates By Exertion And Repetition (Silverstein, 1985)

Studies such as the above provide evidence of the relationship between risk factors such as force and repetition and CTDs. In this case, a study population of 574 employees was classified into four groups, then cases of CTDs were analyzed for each group.

The disorders analyzed included problems with the neck, shoulder, elbow, and hand/wrist. One important finding for CTDs of the wrist is summarized in the above graph. There were cases of CTDs among the first group, which were standardized at 1.0, as shown in the graph. The rates of CTDs for the second and third groups were about three times that of the control group. The rate for the final group was almost 17 times that of the control group.

The implications of this study are:

(1) Either force or repetition alone can increase the risk of a CTD;

(2) There is a *multiplying* effect. This is the bad news—the combination of high force and high repetition was exponential, not linear (This study only evaluated force and repetition. It is quite likely that if other risk factors were included, such as vibration or posture, these differences would be more striking yet.);

(3) Conversely, there is a *dividing* effect. This is the good news. The above graph shows that even if improvements are made to reduce only one of the risk factors, the risk of a CTD drops drastically.

The last conclusion is reassuring. It helps show that even if a job cannot be made perfect, it is possible to help prevent CTDs by addressing just one of the risk factors. Any improvement can be worthwhile—in this business, a little bit can go a long way.

Several considerations are important in understanding these risk factors:

- The supporting evidence that links these factors to CTDs is well documented in the scientific literature. Currently, the American National Standards Institute (ANSI) Z-365 Committee on Cumulative Trauma has compiled probably the best review of studies and data. This committee has reached consensus concerning these risk factors and has adopted a framework for these factors equivalent to the above.

- The levels of exposure—that is, how many repetitions, at what levels of force, and in what posture—that can trigger a disorder are not yet known. Moreover, precise measurement of these factors is often difficult. All that can be said at this point is that more is worse and less is better. (An exception is for certain lifting tasks, where guidelines have been developed.)

- The more factors involved, the greater is the possibility of developing a CTD. Conversely, if any of the factors can be reduced, the risk of a problem can be lessened, even if all of the factors cannot be eliminated completely.

- Not all employees exposed to these factors will be affected. That is why the term *risk* is used—the risk of getting a disorder may be higher or lower depending on the exposure, but it is not inevitable that a problem will result.

WHY CTDs ARE NOT SO MYSTIFYING

In some ways, it is not so mystifying that these disorders occur, especially on highly repetitive, assembly-line-style work. When machines do those types of heavy or repetitive movements, we expect them to fatigue and fail, especially if they are not properly maintained. Indeed, for machines and equipment, we often speak of "mechanical stress points" and "duty cycles to failure."

In many ways, humans are no different. Any moving part in the human can also fatigue and fail, especially if there is insufficient maintenance. It should come as no surprise that equivalent disorders would arise at work from equivalent tasks.

Of course, the human body can repair itself, unlike a machine. But the human body needs sufficient time to recuperate. Unfortunately, the rigors of modern worklife often do not allow time for this rest and recovery.

Cumulative Trauma Is Age-Old

"... the harvest of diseases reaped by certain workers ... [from] irregular motions in unnatural postures of the body."

"... constant writing also considerably fatigues the hand and the whole arm on account of the continual and almost tense tension of the muscles and tendons. I know a man who, by perpetual writing, began first to complain of an excessive weariness of his whole right arm, which could be removed by no medicines, and which was at last succeeded by a perfect palsy of the whole arm. ... He learned to write with his left hand, which was soon thereafter seized with the same disorder."

"I have noticed bakers with swelled hands, and painful, too . . . [from] the constant pressure of kneading the dough . . . "

Bernardino Ramazzini *(considered the founder of occupational medicine)*, Diseases of Workers, *1713*

"In consequence of its often being caused by such movements as wringing clothes, it is known as: washer women's sprain."

"The tendons of the extensor muscles of the thumb are liable to become strained and their sheaths inflamed after excessive exercise, producing a sausage-shaped swelling along the course of the tendon, and giving a peculiar creaking sensation to the finger when the muscle acts."

Henry Gray, Anatomy, Descriptive and Surgical, *1896*

Butcher's Wrist in the 1920s

Old-time meatpackers have referred to this disorder as "butcher's wrist." One retired packinghouse supervisor mentioned, "Yeah, I remember my old man—a ham boner by trade [in the 1920s]. His hands would swell up like a prize fighter. Everyone else had it too, but of course no one would complain. They'd lose their job."

Another old-time packer from rural Kansas, a superintendent by the time the ergonomics movement came to his workplace, stated: "When I first hired on as a kid [in the late 1930s] and I started boning, my hands swelled up and they hurt like hell. I went to the doctor and he diagnosed me as having 'cornshucker's disease.'"

The latter story indicated several things:

1. Meatpacking employees have had problems for years—and these were not just whiners and complainers. Here was someone who ended up in upper management.

2. The incidence of wrist CTDs among agricultural workers must have been so high that it got its own name, at least in that community. The wrist-intensive work of shucking corn by hand was apparently causing an epidemic of problems.

3. The image of a manual farm laborer in rural Kansas in the 1930s is not that of a pampered couch potato. One would consider them to be tough and wiry, and of the noncomplaining sort. Yet, they were clearly having problems, even back then. (This demolishes the argument that the reason why CTDs are appearing today is that our society is turning "soft." If farm laborers in the 1930s could get CTDs, anyone can.)

Even Marine Sergeants in the 1950s

As one becomes more aware of these disorders, it becomes easier to find mention of them in novels and history books. One such example is found in a history

MALINGERING?

The Owner

During a presentation to a group of owners of small meatpacking companies, I was starting to describe carpal tunnel syndrome when one of the participants raised his hands overhead to reveal two fresh scars on his wrists. "I just had surgery for carpal tunnel syndrome," he told everyone. "We've been having a hard time keeping up with orders lately. So I went out to work on the floor. This is what happened." And he owned the company!

The Contest Winner

I received a call from the CEO of a medium-sized company. He informed me that "we need help in addressing CTDs. My secretary just had surgery on both wrists." He continued, "I didn't know she was having any problems until her husband called me. It took both of us together to get her to go to see the doctor. That shocked me, since she's such a hard worker and so fast. *She wins typing contests, you know.*"

The Hard Workers

One of the large meatpacking companies began to instruct their employees to report symptoms, as part of their ergonomics program. They had taken this step with some trepidation, fearing a sudden rush of malingerers. But what happened wasn't what they expected.

"What we got were the hard workers," reported the safety director. "We got the ones who didn't want to say anything, because they didn't want to be perceived as whiners or complainers. But now that they've come in, we can do something for them. These are some of our best people."

The Concert Pianist

The newspaper announced that a world-renowned concert pianist canceled a presumably lucrative tour because of tendinitis.

of the Korean War. This reference is striking because it concerns a Marine sergeant in wartime:

"I found a note from my clerk-typist, Sergeant Shafer, informing me that he had been so badly overworked in the field that he was turning into the hospital with a lame wrist."

S. L. A. Marshall, military historian, referring to
Korea in 1951 (emphasis added)

Space Invader's Wrist

By the 1970s, CTDs were cropping up in increasing numbers. Medical journals had a relevant article or letter in every issue, it seemed, describing some new disorder—"space invader's wrist," "Nintendo thumb," and the like. As of 1979 in the auto industry, few people had heard of carpal tunnel syndrome or tendinitis. But as industry safety and health experts became aware of the problem and investigated a few plants, the problem was found to exist virtually everywhere. By the mid-1980s, especially with the large fines issued by OSHA, most workplaces were noticing the problem.

WHY CUMULATIVE TRAUMA DISORDER RATES ARE INCREASING

There are two main reasons why CTD rates have exploded. The first reason is simply an increase in awareness and subsequent reporting. The second is that there have been subtle changes in technology that have increased exposure to the risk factors for CTDs. Additionally, there have also been suggestions that the increase is related to a decrease in physical fitness in the population, or that it is only a fad. Details on each of these explanations follow.

Increase in Awareness and Reporting

As the examples on the previous pages indicate, CTDs have existed for a long time. In all likelihood, people have been affected by cumulative trauma for centuries, if not since the Stone Age. The paintings and drawings from past centuries show farmers stooping in their fields and fishermen hauling in nets, work conditions that were undoubtedly painful and injurious.

However, these disorders have been recorded only recently. Earlier, people simply had a sore back or a lame wrist and suffered with it. One could argue that if the actual incidence of CTDs and overexertion injuries were known for a century ago, it would have been worse than today's.

At present, there is increased recognition and reporting as a result of a variety of factors:

- As media exposure increases, people hear about CTDs and recognize the symptoms in themselves.
- Both managers and health care providers are beginning to realize that the sooner a disorder is discovered, the better are the chances for treatment. Thus, there is a heightened proactive effort to encourage employees to report problems, such as by holding employee training sessions or including articles in workplace newsletters. Consequently, the number of recorded cases rises.
- Expectations about work appear to have changed. People do not accept the idea that work necessarily involves pain and worn-out bodies, at least in relation to what previous generations would have accepted.
- Employees realize they can sometimes receive compensation for suffering (or feigning) a disorder, so there is a financial incentive to report problems. In contrast, reporting a problem a century ago would likely have been equivalent to volunteering to be laid off.
- Diagnostic techniques have improved and have resulted in greater use of medical terminology for ailments. Moreover, "rotator cuff tendinitis" is more likely to be recorded than a "sore shoulder."
- Companies are expanding their on-site medical capabilities, for example by hiring occupational health nurses. Once there is a nurse on-site, recorded cases of CTDs can increase, simply because of increased awareness and easier access to medical care and treatment.
- Companies are required by OSHA to record work-related disorders, and have been fined for not doing so. Moreover, the criteria for recording have been

CONSCIOUSNESS-RAISING

In the late 1970s, while employed by the UAW in Detroit, I routinely attended meetings of auto workers. As we began to learn more about CTDs, I began to ask, "Have any of you ever had surgery on your wrists?"

Those who had the surgery would raise their hands. Then there would be stunned glances around the room as everyone realized that a good portion of the group had their hands in the air. "I thought it was just me," several people would say simultaneously.

What everyone had thought was a rare affliction based on personal factors turned out to be common among those who did repetitive handwork in auto plants. There was an epidemic under our noses, but no one knew it until then.

changed so that even minor cases of CTDs are now being recorded. Many employers seem to be erring on the side of overreporting, rather than risk a fine.

In some industries, CTD rates in the OSHA recordkeeping system have jumped as much as 300 percent in one year for an entire industry. It would be difficult to explain such a drastic increase by a drastic worsening of conditions. It is more likely that there has been a sudden recognition that disorders need to be reported on OSHA logs.

Changes in Technology

The nature of work has changed subtly through the years in several important ways that increase the risk of CTDs:

Increased Specialization of Tasks

A key shift is that the division of labor has become increasingly narrow. An employee who two decades ago would have done a variety of tasks involving many different work motions, now may be doing only a few tasks that involve the same motion repeatedly.

The classic example is the switch from the manual typewriter to the electronic keyboard. In the old days, you would grab a sheet of paper, stick it in the typewriter, roll it down, bang away in relatively sweeping motions, hit the carriage return, bang away a bit more, and in time need to stop to take out the paper, put it somewhere, grab another sheet, and start over. For mistakes, you would reach for correction fluid and in the process use different motions, plus get a microbreak. If there were carbon copies to be corrected (remember carbon copies?), that would also involve a set of somewhat different motions.

At present, with the electronic keyboard virtually all you do is type. There is no need to change paper or to hit the carriage return. You keep moving your fingers, faster than ever before. Moreover, you can do word processing, bookkeeping, scheduling, drawing, and many other tasks that you could not do on a typewriter, so you end up sitting there for longer periods of time—with endless finger motions. And you can be assigned to the word processing pool, where you may not even get the chance to do filing or answer the phone or any alternate work.

And thus it is not uncommon now to see wrist disorders among people who work at the keyboard. To be sure, typists in the past probably had problems that went unrecorded. But clearly technology has changed in a way that has multiplied the inherent risks. It is not merely an increase in awareness that is behind the increased CTD rates.

As a side note, it is important to emphasize that not everything has gotten worse. With the electronic keyboard, two important issues have improved: (1) there is much less force needed to activate a key and (2) there is less distance traveled by fingers and thus less tendon activity. What has worsened is the repetition, the duration, and the number of people exposed. Keying is faster, and we do more of it for longer periods of time.

The same trend has been experienced by workers in other fields, such as among architects, design engineers, and drafters. Previously, their tasks involved a wide array of motions while working on drafting tables, using pencil on paper. Now this work is done almost completely by computer, using computer-aided design (CAD) systems. The consequence is often excessive use of one wrist to manipulate a mouse or an electronic stylus.

This increased specialization of labor has affected many industries and occupations. Not only can the tasks involve more repetitive motions, but also people may remain in the same posture longer than was the case previously, such as by standing or sitting all day. Although work may have become less strenuous, the exposure to other CTD risk factors has increased.

Increases in Assembly Line Work

A related trend has been changes in the kinds of work done in a high-volume, highly repetitive fashion. Several decades ago, the stereotype of repetitive work was the automobile industry. Now this type of assembly line work is found in many industries.

A good example is the food industry. The increase in demand for prepared foods has meant that new types of workplaces have been constructed to process the food. What was previously done at home intermittently is now done in a workplace continuously. Employees work in a repetitive, assembly line fashion preparing foods to be sold in convenience stores and even gas stations. As a result, people are now exposed to CTD risk factors in ways they had previously not been.

Gaps in Automation

Another way that technology has changed work and contributed to CTDs is a by-product of automation. Many industries have become highly automated in recent decades, affecting the workplace and work tasks in countless ways. But some types of manual work remain when the technology has not been sufficiently refined to automate everything.

This area is often neglected. It is not uncommon to see highly mechanized workplaces in which the scattered gaps in the automation are filled by humans. In these cases, the people often perform very simple and highly repetitive tasks. A typical such task involves pick-and-place motions to transfer products from one automated process to another, often where some visual judgment must be made. Humans are left to fill the breaches until technology develops to automate that aspect as well.

Population Changes?

There is some debate whether the population as a whole is becoming more vulnerable to CTDs. Clearly, there are *societal* issues that have changed, but these are aspects of awareness, as described above—less willingness to work hurt, increased incentive to report problems, and so forth.

But the debate here is a *physical* issue. The argument is that, as a populace, our strength and conditioning is worsening. Many fitness and health experts believe that the U.S. population as a whole is not in as good shape as it was a decade ago. And being in poorer shape may increase susceptibility to CTDs, plus reduce the possibilities for speedy recovery.

While there may be merit to the concept that physical conditioning affects individual cases, the argument that a change in the functional work capacity of the entire population is behind the CTD epidemic is refuted by two crucial facts:

- Athletes, in better shape than the general population, appear to routinely have CTDs. Professional baseball pitchers suffer arm disorders not because they are out of shape, but because their task requirements exceed the capabilities of the human arm.
- If the problem were the population, we would expect no differences in CTD rates in different industries. But CTD rates do vary by industry, so there must be something else afoot. (An alternative—but absurd—explanation would be that the people who are most out of shape choose to work in strenuous industries like meatpacking.)

Fads and Malingerers?

There are some who argue that the whole CTD issue is nothing but a fad that will simply disappear in time. For example, in Australia, CTDs have been labeled the "Australian disease" and viewed as a symptom of national malaise that should be countered by pampering employees less (argued under the assumption by some Australians that CTDs occur only in that country).

To some extent, the case for faddism may have some validity, at least in the following way. The nature of reporting symptoms of CTDs involves awareness and paying attention to how people feel. As the issue of CTDs is highlighted in the press and at work, undoubtedly people report more problems than they would otherwise. Subsequently, in a few years when information on CTDs and ergonomics becomes more routine and prevention programs are put in place, one would expect a drop in reporting simply because of less attention paid to the issues. In this sense, the current attention could have the appearance of a fad.

But to claim that the whole rise in CTDs is a national fad is nonsense. The evidence to the contrary is too strong.

Malingering is a related claim. Unquestionably, some people take advantage of the system when they can. This is especially a problem with medical disorders such as CTDs where the primary presenting symptoms are pain and discomfort that cannot easily be objectively measured or validated. The system relies on taking people at their word, which opens opportunities for abuse.

However, this is not the same as saying that malingering is the primary cause of the increasing incidence of CTDs. Indeed, as companies are now beginning to proactively conduct CTD surveys, a common finding is that there are many people with problems who have never reported them—perhaps two or three times as many people as the official OSHA injury and illness logs indicate. Most CTD victims are hard-working people who, rather than shirking, may have been overdoing it.

THE NURSES WERE STUNNED

In a large company, we arranged to have the plant nurses provide physical exams to *all* employees in one department where a number of employees had reported CTD symptoms. By the time the survey was completed, the nurses were stunned by the number of problems they were finding.

"I thought we'd seen everyone, for all the 'whiners and complainers' we get in here," exclaimed one nurse. "But look at all these people."
Replied another, "I bet for every malingerer that we've been seeing, there's four or five people with problems out there that have never said, 'Boo.'"

As a final thought, if the sudden and widespread rise of reported CTDs is related to increased malingering, then it is a symptom of an even greater social problem. A nation of malingerers would be a sign of underlying malaise and alienation of employees, an insidious rebellion against the system. Thus, those who are skeptical about the notion of CTDs and suggest the real issue is that of widespread malingering, may in fact be making the case that an even greater problem exists.

Why CTD Rates in the Meat and Poultry Industry Are Higher

The meat industry currently has higher rates of CTDs than other industries, although this rate is now decreasing while the rates in other industries are still in-

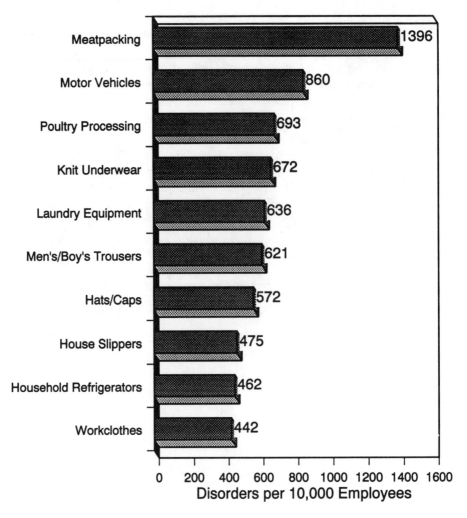

The ten industries with the highest rates of CTDs (upper extremities only; these numbers do not include lower back injuries). Note that all ten industries fall into three categories: food processing, metal work, and textiles. The common denominator of these industries is that a high percentage of the work force in each does repetitive hand and arm work. (*Source:* Bureau of Labor Statistics, 1993.)

creasing. As such, it provides a case study in increasing our understanding of CTDs. Such a study can help managers in all industries gain a perspective on their own operations.

Furthermore, the experience of the meat industry provides a case study in how management can address the CTD problem. This industry was the first to come under regulatory action, and the industry has responded with some of the most comprehensive CTD prevention programs in the United States, in many areas expanding knowledge about how to deal with this issue.

It should be highlighted that the problem of CTDs in the meat industry is not restricted to the United States. From all indications, the same problems exist in other countries with large meat industries, such as Denmark, Canada, Australia, and New Zealand. CTDs in meatpacking are an international problem.

There are a variety of reasons why this is the case. All reflect specific applications of the general trends described above.

Hard and Heavy Work

Throughout history, meatpacking has always been hard and heavy work. Clearly, CTDs existed 50 years ago. They were just not reported; employees either accepted the pain or quit. There is some validity to the argument that if the actual rates of cumulative trauma in the past were known, they would not be much different from today's, and perhaps were even higher.

The Demise of Neighborhood Butcher Shops

The switch to more assembly-line-type work has affected this occupation. In the old days, neighborhood butcher shops did much of the repetitive boning and cutting work. The slaughterhouse merely provided sides of beef and left it to local butchers to do the rest. The butchers themselves would break up the repetitive cutting tasks by waiting on customers and doing other related tasks. They would get intermittent relief from the cutting table.

Today, that work is done in the slaughterhouse along big assembly lines (or to be more precise, along *dis*assembly lines). Hundreds of people work in these plants standing alongside cutting tables doing repetitive handwork with little opportunity for relief by doing alternative work.

Difficulties in Mechanization

Most other industries are more automated than the meat industry (and food industry in general). The standard approaches to automation that are now taken for granted in operations such as car assembly plants cannot be used in meat plants, for two reasons:

1. Meat products (and food products in general) are of various sizes, which makes mechanization difficult. This situation stands in contrast to operations such as car assembly, where the exact shape, size, and location of each item coming down the line are known with precision. Thus, the techniques of automation that were introduced into general manufacturing in the 1950s are still not possible in meat plants.

2. Sanitation requirements in the meat industry make any equipment development much more difficult. It is hard to build equipment free of nooks and crannies and porous materials—the breeding grounds for bacteria. Moreover, equipment in meat plants needs to be washed down regularly with high-pressure hoses. Most equipment routinely used in general manufacturing would not last long in meat plants. Electrical components would short out and metal surfaces would rust; in a meat plant, most equipment must be made of stainless steel.

Consequently, the design constraints for equipment in the meat industry are severe. Opportunities for automation have been few. As a result, a higher percentage of the work force performs manual work in the meat industry than in other industries, which is reflected in the CTD rate.

Multiple Risk Factors

Finally, in contrast to many other industries, employees in the meat industry often are subject to simultaneous exposure to all of the CTD risk factors. And as we have seen, CTD rates rise exponentially when there is exposure to multiple risk factors. Meatpacking tasks involve motions that are both highly repetitive and forceful, done frequently from awkward postures. Moreover, many meat employees often use vibrating tools and work in the cold for their entire shift. This is true throughout the world, not just the United States.

COSTS OF CUMULATIVE TRAUMA

The financial costs can be as severe to employers as the physical symptoms are to those affected. Cumulative trauma can increase costs in a variety of direct and indirect ways:

Workers' Compensation	A growing portion of workers' compensation payments are related to cumulative trauma of the lower back and upper limbs—lost time, medical treatment, and disability costs. With many companies with which I have worked, this portion is about 75 to 85 percent.
Turnover	Dissatisfaction caused by fatigue, working in uncomfortable postures, and experiencing symptoms of cumulative trauma may easily lead to increased employee turnover.
Absenteeism	Similarly, a common reason for worker absenteeism

is that they are experiencing early stages of a cumulative disorder. Work that hurts is not satisfying.

Morale Discomfort, aches and pains caused by poorly designed tools and workstations can affect morale.

Product Defects People working at awkward and uncomfortable workstations or with poorly planned procedures are not in a position to do their jobs "right the first time." Errors are more common.

Production Barriers Jobs that are hard on people are often bottlenecks in production. Inadequate tools and equipment can be inefficient.

Red Tape The paperwork required for repetitive trauma cases can also entail significant staff time and costs.

OSHA Fines Finding alternative work for affected employees can cause considerable disruption in the workplace. Some of the largest fines issued by OSHA have concerned repetitive trauma and related recordkeeping.

Many companies have found that multiplying their direct workers' compensation costs by 4 provides a good estimate of these overall costs (some investigators suggest a 10:1 ratio). For example, in 1992, typical costs for a carpal tunnel surgery case with six weeks of disability were $25,000. Multiplied by 4, the estimated total costs would be $100,000!

CASE STUDY: HIDDEN COSTS FROM SIMPLE NECK PAIN

The executive chef of a small gourmet catering and deli shop experienced neck and shoulder pain. The resulting costs were many, both for the chef and for the shop owners.

The ergonomic issues in the shop were many. The chef was tall and all work surfaces were low, thus requiring the chef to hunch over almost continuously. Periodically, the chef needed to lift heavy loads on and off the stove and from one preparation surface to another. The floors were hard and had no antifatigue mats. There were high heat and cold fluctuations. Stress was also a factor since the chef was under pressure both to get product out and to time food products to be cooked in the correct sequence and end up finished at the same time. And as a good professional, the chef did not merely take a cookbook approach, but judged factors such as humidity and the consistency of raw ingredients in preparation and cooking.

The Christmas and New Year's season involved the heaviest demands and work loads. It was also the most profitable time for catering. Unfortunately, a few days before Christmas, the chef woke up at 5 a.m. with debilitating pain in her neck and was taken by her husband to a hospital emergency room. She stayed home two days, then went back to work, restricted from doing heavy tasks, not just by her physician, but also by physical limitations—she could not move as quickly as before and could not lift half of normal. She was required to wear a neck brace, which substantially restrained her movements. She worked through New Year's to help out her employers, then was forced to quit because of the pain.

Costs Incurred

- The medical costs were incurred through the employer's short-term medical disability insurance program. (The chef did not file a workers' compensation claim.) — $1000
- Two days of lost time were incurred in the busiest, most profitable season. Orders were filled inadequately and some were lost. — $3000
- On returning to work, the chef was restricted from doing normal work. Thus, production was lower than normal. — $1000
- The owners ultimately lost their most highly skilled employee, the one who knew more about the operations than anyone else. Their investment in a key employee was lost and they incurred the costs of hiring and retraining a replacement chef. — $10,000

Total — $15,000

Thus, there were many indirect costs for a simple case of neck pain, all of which were difficult to absorb in a low-margin food operation. None of the costs were captured in the workers' compensation system—all were hidden.

HOW TO PREVENT CTDs

There are two main approaches for management action to reduce CTDs: ergonomics and good medical management programs. When both are integrated into an effective program, the results in decreased injuries and costs can be dramatic.

Ergonomics

To reduce risk factors for CTDs, companies should apply the principles of ergonomics to the design of equipment, tools, and systems. In essence, all tasks should be reviewed systematically and ways sought to reduce exposures to:

- repetitive motions
- exertion
- awkward postures
- contact stress
- vibration and temperature extremes
- stressful work situations

The rest of this book provides examples and more background information about accomplishing the goal of applying ergonomics to solve problems such as CTDs. In particular, the chapters cited address the following specific aspects of CTD prevention: applying ergonomics principles (Chapter Two), minimizing stressful work situations (Chapter Seven), success stories in reducing CTD rates (Chapter Eight), and setting up a workplace ergonomics process (Chapter Nine).

Medical Management

The second way to reduce CTDs is by establishing good medical management systems to identify people at early stages of problems and treat them appropriately before the disorders become serious. It is important for managers to know that CTDs *can* be managed. The evidence is clear that the severity and costs of CTDs can be reduced by an active medical management program, as part of an overall CTD prevention effort. The vast majority of workers treated conservatively can return to usual and customary employment, as long as work modifications are made where necessary.

The medical issues are complex and, for some aspects of diagnosing and treating CTDs, there is little consensus in the medical community as to the best approaches. Nonetheless, CTDs respond to medical management. If the CTD is identified sufficiently early, there is a great deal that can be done to successfully treat people.

Two additional areas where there has been some controversy are exercise and musculoskeletal supports, such as back belts. Both of these concepts may have merit if they are implemented correctly as part of an overall CTD prevention program. However, an important concern from a business standpoint is to ensure that neither approach distracts from the evaluation of work tasks. As long as a company is attempting to make workstation and equipment changes, there is hope that some breakthrough in innovation can be achieved.

More details regarding management perspectives on these medical and related issues are described in Chapter Nine.

CTD RISK FACTORS: SOURCES OF INEFFICIENCIES

Remember, even if CTDs are not a problem in a given workplace, addressing the CTD risk factors can have other payoffs. All of these risk factors can be a source of inefficiencies:

- Excessive motions and exertion often entail wasted time and effort.
- Awkward postures put people at a mechanical disadvantage in performing work.

People who are uncomfortable and fatigued cannot do their jobs right the first time every time.

F IVE

QUALITY AND PRODUCTIVITY

The parallels between ergonomics and the quality improvement process are striking. This chapter discusses these parallels, and outlines the insights that ergonomics provides concerning both quality and productivity. It also makes the point that ergonomics fits well into the efforts to improve quality and to rethink what really increases productivity.

THE ERGONOMICS/QUALITY PARALLEL

There are a number of parallels between concepts of both quality and ergonomics. While ergonomics is a separate and distinct tool in a manager's problem-solving kit, the overlap between quality and ergonomics is important to highlight:

1. Ergonomics fits well into a total quality control process.
2. The concepts, tools, and approaches of the quality process can be used to solve ergonomics issues.
3. Ergonomics can readily be used to improve quality.

The two processes reinforce each other in a variety of ways. Organizations that have a true understanding of quality typically are the ones that most easily grasp the value of good ergonomics. It fits in with all they have learned in recent years.

Parallel Definitions—Meeting Customer (User) Requirements

Quality is often defined as meeting customer expectations and requirements. Ergonomics is essentially defined in the same way—designing tools, systems, and products to meet the needs of the user. In quality jargon the phrase is "know the customer (internal or external)." In ergonomics, this is simply "know the user."

To be sure, the overlap in definition is not 100 percent. Quality can include customer expectations about aesthetics, such as the paint finish on an automobile, or about life expectancy and warranty of a product. Neither of these concerns is a part of ergonomics, since there are no human/tool interactions in those parts of the system. Furthermore, ergonomics emphasizes some issues such as physical fit or operational expectations of which customers are not always consciously aware. Nonetheless, the similarities in basic concepts are striking.

Parallel Problems—Defects (in Products and Humans)

Much has been said and written in the past decade about product defects. The traditional pressure in industry for quantity of output has led to problems in quality. The ergonomics side of the coin is that the same narrow focus on production has led to human "defects" by creating cumulative trauma disorders (CTDs) as well as mental stress-related disorders and human-caused errors.

Concern for quality helps put much of the CTD issue in perspective. The traditional pressure for quantitative output has led to many management practices that not only contribute to quality problems but also increase the risk for CTDs through the same process. In particular, emphasis on working harder and faster, rather than smarter, has increased the exposure to physical risk factors such as forceful repetitions, plus led to management systems that increase the psychological stress on employees (see Chapter Seven). The reverse is also true. The changes in management systems needed to promote product quality can reduce the stressful situations at work that contribute to CTDs.

Thus, a core issue in common between ergonomics and quality is the problem to be overcome—defects. If the only goal of production is to churn out large numbers of products as fast as possible, at least three negative outcomes can result: product defects, human error, and CTDs. These represent major challenges to business as usual as we enter the twenty-first century.

TERMINOLOGY: TOTAL QUALITY MANAGEMENT

The approaches to Total Quality Management vary, as do the schools of thought on different aspects of implementation. Terminology may also vary. The phrases "quality improvement process" or "Total Quality Management" are known in some organizations as "manufacturing excellence," "world-class production," or other equivalents. Nonetheless, there are typically a number of standard elements to quality management. It is within these common elements that ergonomics fits.

Parallel Process—Implementation

The types of management initiatives required to install the quality process in the workplace are the same as those needed to implement sound ergonomics. In fact, some companies use the mechanisms of their quality process to implement ergonomics activities, rather than try to develop a new structure.

The process of quality parallels the process of workplace ergonomics. The following program elements show these overlaps. (See Chapter Nine for more details on ergonomics activities.)

Continuous Improvement

A standard concept of total quality programs is that of continuous improvement. Continuous improvement means that there are no permanent solutions to any particular issue, and that improvements can—and should—always be sought for each aspect of a system. Once a change has been planned and implemented, it should be evaluated to see what further improvements are needed, then those changes planned and implemented. The cycle is never-ending.

Ergonomics also involves continuous improvement. Nothing is "ergonomically correct" in the sense that a problem can be fixed once and for all and then forgotten about. In general, there will always be better ways for every improvement made. Furthermore, ergonomics itself can provide insights and perspectives to help make continuous improvements.

Work Process

Much of the quality message has been: concentrate on the work process, not on the end results. By this, quality experts mean to deemphasize (1) end-inspection of products, and (2) arbitrary financial and production targets set by upper management that are often made with little understanding of their company's production process and its possible shortcomings. Focus on the process, quality experts say, and the results will take care of themselves.

While setting goals is important, the point is to focus on improving the production system as a whole. And ergonomics inherently deals with the work process. By paying attention to ergonomics, the production process can be improved and production goals more easily met.

Management Leadership

Visible management leadership is important for both quality and ergonomics. Commitment and involvement from top management are essential elements to convince employees that both quality and ergonomics are required for the success of the business. Management by example has additional importance here.

Employee Involvement and Empowerment

Another element of most Total Quality Management systems is employee empowerment. New approaches are being developed to provide employees with as much control as possible over the daily events of worklife. Employees gain input and involvement in decision-making in workplace operations. There are countless advantages to distributing responsibility in this manner, but basically management benefits from employees' intimate knowledge of work processes and employees benefit from increased fulfillment of their capabilities. Both parties can share in the financial rewards.

Likewise, workplace ergonomics programs generally need to involve employees in the activities. Employees have insights into their own jobs that can lead to job improvements. Indeed, many issues simply cannot be known by people who have never done the job. Furthermore, the involvement helps develop ownership in a new technique or piece of equipment that promotes acceptance and ultimate success of the change.

As a side note, if an organization is only beginning to contemplate a formal system to involve employees, an ergonomics program is an excellent place to start.

TABLE 5.1   *The Parallels of Ergonomics with One Particular Approach to Quality*

Four Absolutes of Quality	
Requirements	Ergonomics is understanding human requirements
Zero defects	Human well-being
Prevention	Principles of ergonomics
Price of nonconformance	Human errors, defects, injuries, workers' compensation costs, lost time, lost production

The issues and benefits are typically much more immediate than with other issues. Identification of problems and improvements are often quite close at hand and generally any improvements made are felt immediately by the people involved.

Team Building

Quality improvement efforts often emphasize the importance of team effort, as have many workplace ergonomics programs. Experience with both areas has shown the value of an interdisciplinary approach to problem-solving, not to mention consensus-building for implementing changes.

Training

Quality efforts reinforce the need for more training of the work force, both in specific technical skills as well as more general areas such as improving communications skills and group problem-solving skills. Similarly, effective workplace ergonomics programs typically emphasize training in a variety of ways, such as principles of ergonomics, improved work methods, and use of ergonomically improved equipment.

Corrective Action Process

Total quality programs often incorporate a special mechanism for making improvements. Typically a team process is used to: recognize problems, identify root causes of the problems, brainstorm improvements, set action goals, assign responsibility for following through, and evaluate changes.

For ergonomics, the process is equivalent; the same mechanisms can be used.

Some companies form Ergonomics Teams to use this process in addressing ergonomic issues. Other companies simply use their Quality Corrective Action Teams to include ergonomics.

Problem-Solving Tools

Many analytic and problem-solving tools are the same for both areas. Likewise, the emphasis on measurement is equal.

To make ergonomic improvements, these quality tools are needed and work well. Quality experts and engineers trained in these problem-solving techniques usually recognize quite readily the application of these techniques for addressing ergonomic issues. Incidentally, quality experts and engineers would do well to learn more about the growing capabilities in ergonomics for measuring human-related issues and setting human-based criteria for design.

The following list outlines these standard tools and their application in ergonomics:

Quality Tool	Ergonomic Application
Statistical process control (SPC) charts	Understanding the human role in creating variation.
Pareto analysis	Setting priorities for addressing cumulative trauma, human-related errors, etc.
Price of nonconformance (PONC), cost of failure, etc.	Workers' compensation costs, other human resource costs such as turnover and absenteeism, human error costs, etc.
Cause and effect (fishbone) diagram	Identifying causes of ergonomic problems.
Root-cause analysis	Identifying causes of ergonomic problems.
Brainstorming	Identifying options for ergonomic improvement in a team setting.
Flow diagrams	Understanding the work process from an ergonomic perspective.
Various statistical techniques	Measuring ergonomic improvements and various other analyses and evaluations.

JUST-IN-TIME AND ERGONOMICS

The relationship between just-in-time (JIT) manufacturing and ergonomics shows a number of the inherent overlaps—and some issues that increase the importance of ergonomics.

For example, consider a manufacturing facility that switches its internal system of supplying parts to workers. Supplying materials in smaller lot sizes, such as in small tote boxes several times a day (a JIT system) rather than in huge bins full of thousands of parts to be worked on over the course of several days, provides two inherent ergonomic advantages:

• Small tote boxes can be easier to move and lift, with less strain on the arms and lower back.
• Small tote boxes typically involve fewer long reaches and eliminate the all-too-common wasted time and effort to stand up and reach down into a large bin.

On the other hand, some JIT improvements may require additional types of ergonomic interventions. For example, switching to quick changeover systems often means that a task previously done only once a week is now done several times per day. If that task is awkward or hard to do, but not previously excessive because it was done infrequently, it may now be a problem under the new system. It would thus require an ergonomics evaluation and some improvements.

Simple Ideas Yield Good Results

As a final note, only a few simple ideas and principles comprise both the quality process and ergonomics. Yet, if these ideas and principles are applied systematically throughout a work process, the results can be dramatic.

Application

These parallels between ergonomics and quality are important for managers to understand for several reasons:

1. Organizations that have invested heavily in the quality improvement process in recent years need to understand that ergonomics is not necessarily a completely

new program. Indeed, ergonomics can be seen as an augmentation to the quality process, perhaps even the other side of the quality coin.

2. Managers need to convey this strong overlap between quality and ergonomics to the organization. Otherwise, ergonomics could be perceived as yet another new program, an appendage that does not fit in. Pointing out these parallels helps keep ergonomics initiatives from appearing disjointed to workplace personnel.

3. In broad terms, the guidelines for preventing cumulative trauma are found in modern quality management. No new systems or new approaches to management need be invented—only new understanding.

4. For Total Quality Management, there is a concomitant need for a quality work environment, quality equipment, and quality people. Ergonomics is vital to achieve this.

5. Finally, the ergonomics effort can take advantage of the momentum gained from the quality process, and conversely, the ergonomics process reinforces the quality initiatives.

A REDEFINITION OF CTDs: NOT MEETING HUMAN REQUIREMENTS

The quality process includes meeting the physical and psychological requirements of "internal customers," that is, employees. The human body has a number of specific requirements, among these certain anatomical and physiological limitations. For example, the body is limited in its ability to tolerate repetitive and forceful work motions without sufficient rest and recuperation. When these limitations are exceeded, CTDs can result. While the precise limits (human requirements) are not known, and may vary among individuals, it is clear that these human limitations do exist. To truly implement the quality improvement process, it is crucial to understand human capabilities and limitations, then design mechanical and organizational systems to meet these human requirements. This field of study is ergonomics.

HUMAN ROLE IN CREATING DEFECTS

Poor Workstations as a Barrier to Quality

Poor design of workstations, tools, and equipment can easily be a barrier to quality, but is often overlooked. People who work in awkward and uncomfortable postures, who are fatigued, or who cannot physically reach the materials they need

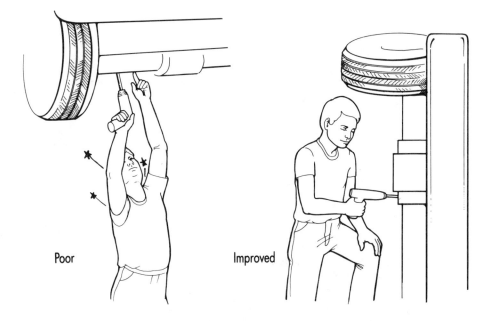

Poor   Improved

The illustration on the left shows a problem common in automobile plants in the past—doing the underbody assembly. Employees had to work in pits underneath the assembly lines, often looking up and working with their arms stretched overhead for long periods of time. The human body does not tolerate working in this posture well for extended periods of time. Fatigue results, with pain in the neck and shoulders, and much difficulty for employees.

Quality can suffer as well. Some employees may not physically reach far enough to do the job correctly. Almost anyone doing this task would become so fatigued that it would be difficult to do their job right the first time every time. This task is designed for failure.

The illustration on the right shows a concept that Volvo pioneered—flipping the car on its side. Here the pit is eliminated and employees work on the shop floor. They are able to use good body mechanics and work in natural postures. And they are in a much better position to do their jobs right the first time every time. This workstation, in contrast to the previous one, is designed for success.

The concept of using powered fixtures to present the product in good orientations to employees is one that can be generalized to many industries. The meatpacking case study on ham boning in Chapter Eight offers another example.

are in no position to do their jobs right the first time. No matter how much time we spend promoting quality, some employees will not be able to achieve quality requirements simply because their jobs are designed for failure.

With ergonomics we can design for success. The fact that the Malcolm Baldridge Quality Award includes recognition of ergonomics as part of the quality effort lends credence to this argument.

Quality experts focus on improving the system and avoid blaming employees.

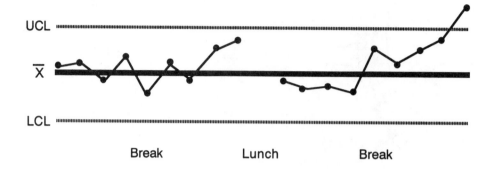

Break   Lunch   Break

> Human fatigue creates excess variation. A manufacturing company routinely measured many aspects of its operations with statistical process control (SPC) charts. In one critical task, variation exceeded normal limits in a way that turned out to be caused by poor ergonomics.
>
> As the charting process began and results were evaluated, engineers realized that what they were plotting in effect was operator fatigue. The task involved standing for long periods of time on a concrete surface with outstretched arms. As the employees tired through their shifts, the variations became out of control. On recognizing this, good ergonomic principles were then applied. The costs of improvement were negligible compared with the cost of the earlier excessive variation.

They assume that people want to do a good job, but the system holds employees back. The ergonomist shares this perspective. Humans make mistakes, but we can be led into those mistakes by poorly designed workplaces. A few examples of how the lack of good ergonomics can interfere with quality production include:

- Poor layout and design of physical workstations can result in unnecessary work, and make it impossible for employees to do their jobs correctly.
- Inadequate workstation design can make it difficult for employees to reach their tools, access parts or products, or see the displays they need to see.
- Exceeding human capabilities may cause fatigue or mental overload, which can result in errors and inconsistent work.

Poor Displays and Controls Promote Errors

Confusing signs, misleading directions, and vague layout of switches and controls can also contribute to product defects. This is one of the most well-studied areas of ergonomics, but an area too often overlooked in the current managerial focus on quality.

- Uninterpretable control panels can easily lead to errors.

- Misleading instructions, whether written or verbal, are a common source of misunderstandings, rework, and wasted time.
- Poor layout and design of workstations can result in mistakes and inefficiencies.

PRODUCTIVITY

There are three principles to learn from this section:

1. Speed alone should not be the focus for improving productivity.
2. Critics of industry should not focus on reducing speed as the only solution to CTDs.
3. A real understanding of how to improve productivity can result in increased output while at the same time reducing CTDs.

It is crucial to understand what productivity actually is. In its most basic form, productivity means getting the most output for the least amount of input. This is an important goal that is consistent with all of the principles and objectives of ergonomics. Indeed, getting more for less has been the basis of the increased standard of living for the past 40,000 years.

Unfortunately, we often behave as though we believe that productivity simply means working harder and faster, in other words, simply increasing input. This misconception causes problems for both managers and critics of industry alike. Often there is a negative connotation to the term *productivity,* based on this misunderstanding and how it is applied.

However, if productivity is thought of as "working smarter," then it can lead to innovation, employee well-being, and increased efficiency. Although the phrase "work smarter, not harder" is becoming a cliché, it does capture ergonomics concepts nicely and is worth reflecting on when doing almost any task.

THE FALLACY OF SPEED (PART I)

Speed is not synonymous with productivity. Getting the product out the door as fast as possible can be a recipe for production disaster:

- Product quality can deteriorate.
- The well-being of employees can worsen.
- The efficiency of the whole production system can suffer from a shortsighted focus on speed.

A related problem is exceeding the production capacity of a physical plant. A plant might have been built to produce 5000 products a day, but now produces

10,000 per day. This causes production problems that are also root causes behind the physical issues confronted by employees:

- Workstations are congested.
- There is no physical room to make any improvements.
- Equipment wears out quickly, often increasing difficulties for production employees, not to mention the sanity, health, and safety of maintenance personnel.

Fortunately, in the context of the quality improvement process, speed becomes less important. System efficiencies are to be maximized, rather than the individual worker's speed. Production quality is the key, rather than the narrow emphasis on getting product out the door.

Much of this message is not particularly new. However, because of the importance of the cumulative trauma issue, and because of the special insights that ergonomics provides on production, the problem of production speed needs to be highlighted.

The fallacy of focusing on increased speed as the primary force to improve production is illustrated by the following items.

A Case Study

Efficiency versus Spinning Your Wheels

The owner of a small manufacturing company and I were watching his employees at work. They were busy hustling about, moving back and forth, picking things up and putting things down. He beamed proudly, "These people know how to work. They're not doing this because I'm here watching them. This is how they work all the time."

But I saw a different picture—inefficiency everywhere. For example, the first task in the production process was to sort incoming raw materials into seven different types. But the work surface provided space for only five. The remaining two types were simply tossed on the floor, where someone else had to pick them up and re-sort them. It certainly looked like a lot of action, but much of it was nonproductive.

The owner was confusing input with output, and speed of activity with productivity. Simultaneous with this inefficiency, the company was suffering from tremendous workers' compensation losses. Employees were experiencing cumulative trauma of the lower back and upper arms from all the moving around, picking up, and putting down. There was little wonder why this was so.

A Scientific Experiment

The graph on the opposite page can be used to make statements on two levels that relate to the "work smarter not harder" principle. First, on an elementary level,

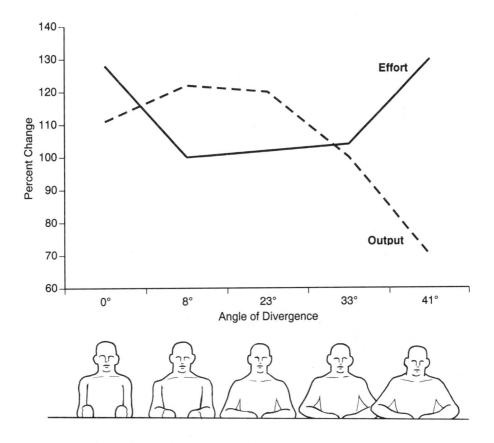

| Tichauer's elbows (1978).

the graph tells us something about good working posture. On a broader level, the same information can be used to help us better understand productivity.

The graph specifically shows the effect of elbow posture on both output and the effort required to do the work. The center of the graph shows the elbows in their neutral posture (8° to 23° away from vertical) which yields the most output for the least input. At the extremes of the graph, the elbows are held either too close to the body or too far away, and consequently performance decreases and the effort required to do the job actually increases.

The first point of the graph merely confirms what seems intuitive. Work is best performed when the elbows remain in their neutral positions at the sides of the body.

The second point, however, goes against our intuition. We tend to think about productivity as meaning the harder we work, the more we produce. This graph tells us the opposite. Here, *the easier the work is designed to be, the greater the output.*

A Microeconomic Model

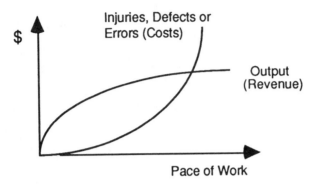

| The cost associated with speeding up the pace of work eventually exceeds the revenue.

This graph shows some of the relationships between traditional concepts of productivity—working harder and faster—and output and costs.

The revenue curve is standard economics. Output increases rapidly with an initial increase in the pace of work, but after a time there are diminishing returns.

Simultaneously, as the pace of work increases, there is a slow increase in associated costs, such as injuries, defects, and errors. As the pace of work increases, the costs eventually rise dramatically.

At some point the curves intersect and thus cost exceeds revenue. It may be difficult in real life to know exactly where they cross, but cross they do.

A Quiz

Productivity Exercise

Try to think of a single example throughout the entire course of human history where any leap in productivity has come from working harder and faster.

_____

Now think of examples where leaps in productivity have resulted from working smarter.

_____

_____

_____

_____

Answer

There have been incremental increases in productivity from working harder with any given technology throughout history, but never any leaps. All major improvements in productivity have come from working smarter.

Indeed, most significant leaps come from improvements that can easily be characterized as ergonomic—taking advantage of human capability or overcoming a human limitation through better organization or equipment.

Productivity is increased not by pushing people to their limits, but by overcoming human limitations through better design.

A Convert

The Time-Study Man

An industrial engineer, after having been involved in an ergonomics program for about a year, told me "I used to walk around trying to spot where I could eliminate people. The employees would actually boo and hiss when I walked by. Now I try to find ways to design things better for the employees. And now the people come up to me to tell me their ideas. It sure feels a lot better."

A Concept

The Window of Time

The concept of the "window of time" helps shed light on many speed issues. For example, a manufacturing plant had improved operations and mechanized production through the years to the point of more than doubling output—from 7 pieces per minute on an assembly line to a current rate of 15 pieces per minute.

Managers were in a quandary about one particular task, which previously one employee could do well at 7 pieces per minute. Now *three* employees were used along the assembly line for the same task to achieve the 15 piece per minute rate. The number of pieces per employee was thus reduced to 5, yet the task seemed more difficult.

In other words, in the past the employee was given 8.6 seconds to complete the task and the job seemed satisfactory. Now, however, each employee was given 12 seconds to complete the task as it moved along the assembly line. Despite the increase in time available and reduction in work load, the employees complained and seemed to be suffering from increased CTDs.

The explanation was simple. The difference is that the window of time available to do the task was reduced. The products were moving along an assembly line that had more than doubled in speed and employees were in effect being required to hit a moving target. The actual time available to them to complete the task was only 6 seconds, since the products were moving so fast they were beyond reach during the remaining 6 seconds.

Improvements

There are three options for increasing the window of time:

1. *Slow the line.* This results in lower output and would generally be unacceptable in most situations, unless cost reductions from fewer product defects and fewer CTDs could justify the reduced output.
2. *Diverge the line.* The assembly line can automatically branch off into split lines. This takes more equipment and space, but does not reduce output. The same volume of product flows through, but at a slower pace.
3. *Reduce product spacing.* This is an elusive concept, but it works. If you place the products closer together along the line and *then* slow down the line, output can remain the same, but the products move by at a slower pace. Sometimes it takes lengthening an assembly line and spreading employees out a bit, but it works to increase the window of time. Thus, the tasks improve for employees without a sacrifice in output.

An Observation

The Lost Art of Methods Engineering

While I was present in a manufacturing facility, I observed an employee sitting at a workbench, assembling a small product. To his immediate left stood a large square box holding the parts he was to assemble. To his immediate right was a similar box into which he placed the finished product.

He reached continually into these boxes, which was a problem, since both boxes were about waist high and over an arm's length wide and long. It was impossible for him to sit at his bench and reach what he needed. His solution, therefore, was to initiate the following sequence every few minutes:

1. grab a tote pan
2. stand up
3. lean over the box on the left
4. **obtain the needed parts**
5. fill up the tote pan with parts
6. sit down
7. unload the tote pan
8. **complete his task on the product**
9. reload the tote pan

10. stand up
11. lean over the box on the right
12. unload the tote pan
13. **place the completed product in position**
14. start the process over again

Steps 4, 8, and 13 were the only ones essential to the job. Everything else was wasted effort that happened to involve long reaches, awkward postures, and repetitive motions. It was inefficient and was contributing to wear and tear on the employee.

As we began to address possible improvements (smaller boxes, parts stands, and so forth), I discovered that even with all of the inefficiencies, the employee was meeting the time standard for the task. To my shock, I realized that the job had actually been timed to be done in that awkward fashion. Some engineer had studied this task, but failed to recognize the needless motions. Not only would this task make a back pain expert wince because of the long reaches and bending, it would also make Frederick Taylor (the father of time study) roll over in his grave because of the wasted time.

But the point of the story is this: Taylor actually invented something called *time and motion study,* but in too many workplaces this analysis has degenerated into time study alone. We even refer to it mostly in this way—time study. The motion part has been lost. We do not sufficiently address how to smooth the work motions and improve efficient movements as was once taught. Too often, we just time the job as part of a system to get people to work as fast as possible regardless of the job setup. Fortunately, times are changing with the interest in ergonomics, quality, and world-class manufacturing techniques, but there is not yet consistent emphasis on simple work method analysis and improvement.

THE FALLACY OF SPEED (PART II)

Preventing CTDs Does Not Necessarily Mean Slowing Down Production

Faster line speeds have contributed to our present problems, but it does not follow that simply slowing everything down is the ultimate solution. Simply telling industry to slow operations down is not much of a recommendation.

To be sure, in the context of Total Quality Management, it is possible that reduced speed can indeed lead to an increase in output. The number of injuries, errors, and defects can also drop, resulting in a net increase in efficiency.

However, there are many alternatives to simply slowing operations down. The following is a case study that makes this point.

Case Study—How Ergonomics Can Reduce Repetitions but Increase Output

There are thousands of examples of how better design can simultaneously increase output while decreasing risk factors for CTDs. Using machines or power tools is the most common type of example. If a machine or power tool does the work, it does not really matter how fast the process is from a human stress point of view. Manual repetitions and force can be reduced while output is increased.

In the following case, one company instituted an ergonomics program and proceeded to investigate jobs and brainstorm improvements. In the process, they helped develop a device to feed paper products into a type of printing press.

The traditional way was manual, using the left hand to feed the press and the right one to remove the product, flip it over, and place it on a drying table. With the old method, there was one piece produced for each hand repetition. Although the manual method was traditional and unchallenged, it had been causing arm and wrist CTDs among the printer operators for years, including burning and tingling sensations in their hands at the end of a shift. No one had complained about this problem, but the ergonomics committee uncovered it during their job reviews.

The new device eliminates the repetitive work for both left and right hands. The operator was freed to concentrate on preparing for the next production run and checking quality from time to time (involving minimal repetitive hand activity).

The device simultaneously reduced the number of hand repetitions done daily from 5000 to essentially zero, while tripling output as shown in the graphs.

In short, critics of industry should not be so quick to recommend a slowing of operations. And managers should not fear a loss of output by reducing CTD risk

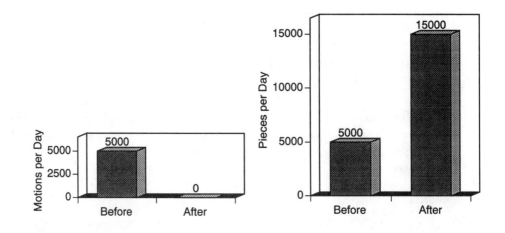

❙ Hand repetitions decreased, but output tripled.

factors. With innovation and good application of ergonomics, repetitive motions and other CTD risk factors can be reduced in a way that increases productivity.

REPETITIVE MOTIONS ARE A WASTE OF TIME

A corollary to the above case history, and perhaps the real message to industry, is that highly repetitive tasks can be time wasters. Once the employee was freed from the task of flipping the product all day long, he could use his time more wisely and more productively.

Use "Work Load," Not "Productivity," When Making Comparisons

Similarly, critics of industry have argued that since productivity has risen so much in recent decades, it must follow that individual employees must surely be more overworked than ever before. However, this theory ignores the effect of improved technology, such as increased mechanization and better layout.

To make valid comparisons of the physical risks and demands that people are subjected to, there is a need to compare tasks by using precise task analysis, for example, measuring changes in work motions per hour, exertion used, and from what posture. Here is where the measurement tools of ergonomics can be used to great effect.

Conclusion: Win–Win Production through Ergonomics

The approach used by ergonomists for improving productivity is straightforward. Instead of using the techniques of (a) always seeking to do a job with fewer employees or (b) always adding more work to a given number of employees, ergonomists adopt the strategy of seeking to do the job with the least *effort* of employees.

In essence, this means systematically applying the principles of ergonomics to all tasks. This approach can only benefit business. If viewed in microeconomic terms, the slope of the revenue curve can be increased (more output for less input, hence greater productivity). Furthermore, the cost curve can be reduced (fewer injuries, defects, and costs for the same pace of work). Hence, profitability can be increased in two ways: higher revenue and lower costs.

Ergonomics can provide considerable insight into the real meaning of productivity, benefiting employees and employers alike. Understanding productivity correctly is particularly important in the debate about cumulative trauma.

If productivity is thought of in a narrow fashion, it can lead to CTDs and product defects. However, if you understand productivity in its broader and more comprehensive sense, you can apply the principle of "work smarter, not harder" to the benefit of the organization and all of the people who make it an excellent place to work.

Sᵢₓ

STRATEGIC ISSUES AND TRENDS

This chapter outlines five strategic human resource trends and several regulatory changes that may affect the profitability of many companies. The value of ergonomics as a vehicle to help management successfully meet these challenges is explained. Both employers and the producers of equipment used in the workplace need to understand these changes in order to prepare for them.

HUMAN RESOURCE TRENDS

Trend No. 1: Higher Workers' Compensation Costs

Most employers recognize that workers' compensation costs are rising. Total workers' compensation costs per employee have been increasing for the past four decades in the United States, as have these costs as a percentage of payroll.

Unfortunately, many managers either accept these spiraling costs as a part of the fixed overhead of doing business, or assume that the only way to seek relief is through their state legislatures. Employers must learn that they have many options for controlling these costs directly, including preventing workplace injuries from occurring in the first place through good ergonomics programs.

Awareness of these rising costs can be used to good effect when justifying purchase of new equipment or renovations of work areas. Often the addition of these costs into the cost/benefit equation can tip the scales in favor of making the improvements (see Chapter Eight, Case Study 3).

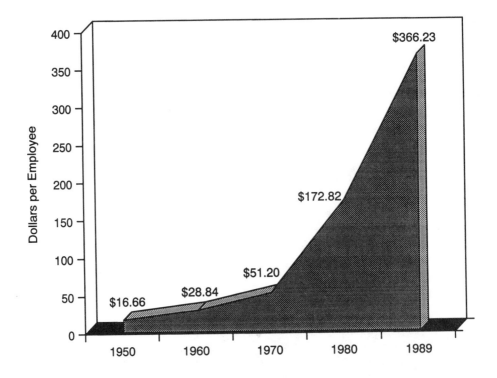

Cost of workers' compensation per U.S. employee for the past four decades. (*Source:* Social Security Administration, 1991.)

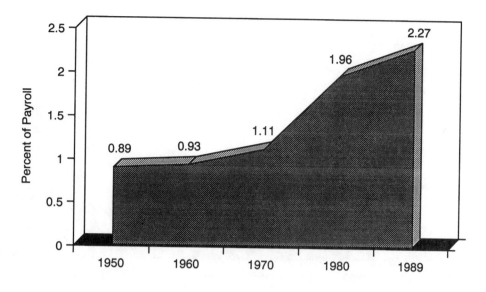

| Cost as a percentage of payroll. (*Source:* Social Security Administration, 1991.)

While this message is aimed at employers, it is also true that employees and unions need to be concerned. This need becomes particularly evident from the above graph, since as workers' compensation costs take a larger share of the payroll, less money for wages and benefits is available for employees.

The rising costs of workers' compensation have become a hot political issue in many, if not all, states. Business groups have been pitted against labor organizations in bitter political fights, with elected representatives and political parties caught in the middle. The media has covered these debates, all focused on the change benefit levels or revamping administrative systems. Unfortunately, seldom if ever has the concept of prevention been raised.

A successful effort to cut the rising costs of workers' compensation simply must include ergonomics. If all employers are as successful in reducing their workers' compensation costs as were the companies described in the case studies in this book, the total burden of these costs on the economy can be drastically cut. Indeed, the magnitude of cost reductions that can be achieved by decreasing benefits or addressing legislative issues pales in comparison with those achievable by setting up good ergonomics programs.

Furthermore, the devisive political issues can be nicely circumvented in a way that makes good sense for the economy. A lose–lose situation can be changed to a win–win one.

There are major benefits for business that will come with a federal OSHA ergonomics standard. (This, of course, assumes a standard that does not micromanage ergonomics programs and that business implements the standards along the lines suggested in this book.)

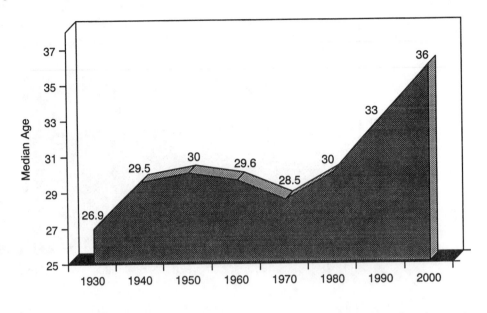

The median age of the U.S. population, with projections for the next decade. (*Source:* U.S. Bureau of the Census, 1990.)

Trend No. 2: Aging Work Force

The American work force is aging. A larger percentage of the working population is older than ever before. This trend is a result of both (a) the baby boom moving into middle age and (b) better medical care and longer life expectancy.

While there are many advantages to having a highly experienced work force, the concern here is for reduced physical capabilities. Although advancing age can obviously restrict the abilities of many employees, applying the principles of ergonomics can compensate for this. Examples include:

Effect of Aging	**Compensating Principle of Ergonomics**
• Poorer eyesight	• Improve illumination and clarity of signs, displays, etc.
• Reduced strength	• Reduce exertion requirements
• Less agility	• Improve heights and reaches
• Increased susceptibility to some types of cumulative trauma	• Reduce repetitions, awkward postures, and exertion requirements
• Decreased memory capacity	• Add visual cues

Good programs in ergonomics are needed to counteract the increased limitations of aging employees. Older employees have more experience, tend to be more reliable, and are trained and educated. When ergonomic adaptations are made, older workers can easily be as productive as younger workers, if not more so.

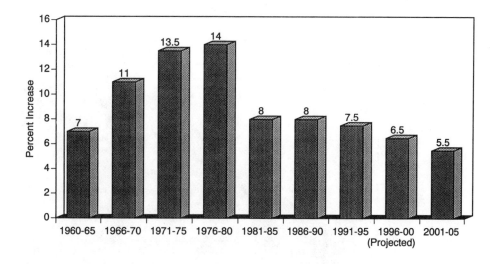

| The rates of growth in the U.S. labor pool. (*Source:* Monthly Labor Review, 1991.)

Trend No. 3: Slower Growth in the Labor Force

Simultaneously, the size of the labor pool is not growing as rapidly as in previous decades. The size of the work force increased from the 1960s until the 1980s as the baby boomers entered the labor force. After the 1980s, the rate of increase in the labor pool began to drop, since new workers are now being drawn from the smaller group—the "baby bust"—born between 1965 and 1979.

These changes affect the workplace in two ways that can be compensated for with good ergonomics:

1. Some employers have in the past relied on a steady stream of young males entering the work force to do physically demanding jobs. This option may not be available in the future, if it is even now. Creative application of ergonomics can reduce the physical job requirements in a way that keeps efficiency high.
2. Similarly, if working conditions are unpleasant, employers may not be able to attract quality employees. Once again, ergonomic adaptations can improve the work environment to respond to this changing labor market.

Low unemployment that comes with a good economy may also require employers to make changes merely to attract a work force. In the 1980s, unemployment dropped in several states in the Great Plains to the point that some companies had to close second shifts—a costly step—because they could not attract enough qualified workers. The same situation occurred in Scandinavia in the 1970s, when unemployment dropped below 1 percent. Employers in those countries responded by investing heavily in ergonomics (MacLeod, 1984).

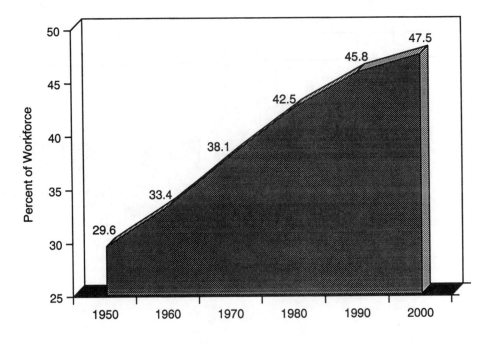

The increase in the female share of the U.S. work force. (*Source:* U.S. Bureau of Labor Statistics, *Handbook of Labor Statistics,* 1985.)

Trend No. 4: More Women in the Work Force

The increase in the percentage of women in the workplace has already made its impact for the most part, but is included here to emphasize the need to modify equipment and tools for women, who are often smaller-statured.

It should be emphasized that generalizations based on gender have not proven to be of much value. Capabilities of individuals are what matters. In the case of strength, for example, we often think in sweeping terms of men being stronger than women. We tend to forget that many women are clearly stronger than many men. Thus, for individual women, no particular modifications need be made, while accommodations may be required for those men who are smaller-statured.

Nonetheless, women tend to be smaller-statured than men. Ergonomists often hear complaints from women that tools and equipment are too large for them to do their jobs well. Thus, ergonomic modifications are needed to enable many women to perform to their full capacity. For example:

- Tool grips should be adjustable or come in several sizes.
- Workbenches should be adjustable, or have stands available.
- Long reaches should be reduced.
- Heavy exertion requirements of many tasks should be improved.

Trend No. 5: Increasing Education and Expectations

A larger percentage of the work force than ever before has completed high school and college. While this trend is hardly news, the actual numbers can be striking.

Although the adequacy of this education has been challenged in the past few years by the scores of American students on international exams, the fact remains that the work force is more educated than even a decade ago. And with this increase in education has come a host of changed attitudes about work.

Today's work force arguably has a higher set of expectations about work than previous generations. It is probably fair to say that our grandparents and the generations before them expected work to be somewhat unpleasant and grueling. People today do not appear to accept the prospects of coming home at the end of the day worn-out and hurting. The comforts and standards of the home environment have improved in recent decades and one can anticipate that most people would expect a parallel improvement in the work environment.

Again, this is hardly a startling revelation. But highlighting this shift reinforces the value of ergonomics. By applying the principles of ergonomics to all tasks, we can design the workplace to help meet current expectations of comfort and ease.

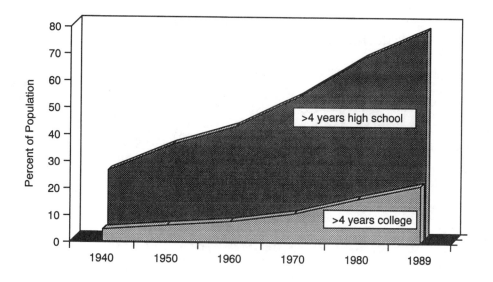

The percentages of the U.S. population with (1) more than four years of high school and (2) more than four years of college. (*Source:* Statistical Abstract of the United States, 1991.)

TWO POINTS TO REMEMBER

Ergonomics Off-the-Job

The preceding sections were written from the perspective of the workplace. However, issues such as aging and higher expectations have equal application in the design of tools and equipment in any part of life. Manufacturers of consumer products need to take note.

Innovation

Whenever jobs are evaluated from an ergonomic perspective, whether to meet any changing human resource trend or for any other purpose, there is always the possibility of coming up with a new and better way of doing the job.

REGULATORY ISSUES

Managers also need to be aware of the regulatory issues that are beginning to affect the workplace. There are several major developments that heighten the importance of ergonomics.

OSHA's *Ergonomics Guidelines for the Meatpacking Industry*

The meatpacking industry has the highest rates of CTDs and, as such, has been the focus of much regulatory and media attention. In the late 1980s, OSHA fined a number of meat and poultry companies in a series of highly publicized actions.

In 1990, as part of an overall special emphasis program, OSHA developed a detailed set of guidelines to help the meat industry establish programs to prevent these disorders. The guidelines were developed in conjunction with the industry in a voluntary and cooperative framework. The industry itself has since begun to establish comprehensive programs in ergonomics.

Implications for Other Industries

This industrywide effort in meatpacking serves as a model for all industries, since the issue of cumulative trauma is new to both industry and government agencies like OSHA. Currently, the meatpacking companies are in many ways a testing ground, determining what types of approaches work best to reduce rates of CTDs.

| The meatpacking industry is serving in many ways as the testing ground to determine effective approaches for preventing cumulative trauma disorders.

The meatpacking ergonomics guidelines themselves are a likely indicator of what OSHA ultimately will expect of all employers. Essentially, these guidelines are administrative in nature, that is, they recommend a management program to deal with the issues, rather than specify design criteria for equipment such as knives or cutting tables.

The guidelines encourage management commitment as evidenced by a written program, employee involvement, and regular program review and evaluation. They are further divided into four program elements:

   I. Worksite Analysis
  II. Hazard Prevention and Control
 III. Medical Management
 IV. Training

A value of these guidelines is that for the first time they officially outlined the elements of a workplace ergonomics program. Chapter Nine of this book contains more information on how to establish workplace activities that meet the goals of these guidelines.

Interpretations

During the meetings between OSHA and the American Meat Institute (AMI) to develop these guidelines (in which the author participated, as a consultant to the AMI), OSHA laid out their expectations in two important ways:

1. Most important was the overall aim of the guidelines, and
2. The details of the guidelines were *not* items that every company had to institute in all minute particulars.

The primary request of OSHA was that management develop good faith efforts to address the CTD issues and implement certain key activities:

- committees formed to coordinate the program
- employees involved in the effort
- training in ergonomics and prevention of CTDs provided for management and employees alike
- jobs systematically reviewed and improvements made where possible
- medical management programs reviewed and improved where necessary

There are many ways of accomplishing the above. Officials of OSHA used the term *menu* to describe aspects of each of the above key elements, stating they were ideas to consider and choose from as each company considered appropriate.

OSHA made it clear in these meetings with the industry that the guidelines were intended to be adapted to the needs and resources of each employer. That is, OSHA expected more activities from employers that were large or had high rates of cumulative trauma, than from employers that were small or had low cumulative trauma rates. Specifically, OSHA officials stated that if an employer had few or no cases of cumulative trauma (assuming accurate recordkeeping), the employer need not do anything at all.

My own view is that both the meat industry and OSHA reacted responsibly and appropriately to the problems in the industry. By mutual consent, rather than the force of law, the guidelines were drafted. OSHA suggested several broad program elements and the industry responded. Both sides avoided getting bogged down in fine legalities or arguing over minutiae. As more experience is gained, a greater understanding has emerged concerning the details of implementing a CTD prevention program—what types of approaches work best under which circumstances.

I personally believe the industry needs to apply ergonomics regardless of whether they have problems with cumulative trauma or not. The meat industry is on its way to planning and designing the packing plants of the twenty-first century. The principles of ergonomics will help the industry achieve that goal.

Federal OSHA Enforcement Activities

Megafines

OSHA has cited and fined employers for cumulative trauma in a variety of industries. Some of the largest fines ever issued by OSHA have concerned cumulative trauma and related recordkeeping.

As of this writing, there are no federal standards concerning cumulative trauma. The basis for the citations has been the General Duty Clause of the federal Occupational Safety and Health Act—the section of the law that states that every employer must provide a workplace free of recognized safety hazards.

This practice has been challenged in the courts, with the legal argument centered around a number of issues such as the extent to which employers must experiment to develop new tools and equipment. Nonetheless, OSHA has successfully cited companies and negotiated agreements that required these employers to address specifically cited jobs and to establish administrative programs to prevent CTDs. These settlement agreements parallel the general elements of the meatpacking guidelines. Indeed, the guidelines themselves were based on the initial settlement agreements from the meat companies first cited by OSHA.

OSHA has invoked the General Duty Clause when inspectors have uncovered what they considered excessive rates of CTDs in a particular workplace. The number of CTDs in each company's OSHA log for workplace injuries and illnesses is the focus of all enforcement activities. At present, if there are no CTDs and the records have been kept accurately, OSHA cannot cite.

Proposed General Industry Regulations

OSHA intends to promulgate regulations for general industry. For the most part, it is likely that the general industry regulations will be based on the meatpacking guidelines; that is, they will require management programs to address the issues, rather than specify design criteria for tools and equipment. The rules-making process will undoubtedly take years, but prudent managers should be aware of the elements of the guidelines, take a proactive approach, and start implementing ergonomics programs as described in the final section of this book.

State and Local Regulation

Several states have proposed either regulations or legislation for general industry concerning ergonomics and the prevention of CTDs. Once again, most of these proposed actions would require establishing administrative programs. At this writing, California appears likely to be the first state to proceed.

REGULATIONS AND GOOD BUSINESS SENSE—IT'S ALL IN HOW YOU DO IT

Regulations should be applied in a way that makes good business sense. While this statement ought to be self-evident, it is not always so. Shortsightedness can be an affliction of both regulators *and* employers themselves. When rules are followed by rote, regulations can seem more burdensome than need be. Two comments are in order:

Be Creative

For ergonomic issues, you can respond to the impending OSHA requirements either in a plodding, bureaucratic way, or you can understand the end goals and be creative on how to get there. Ergonomics is an engaging field, alive with new opportunities for all parties involved, whether you are a regulator or one of the regulated.

Ergonomics regulations can appear daunting, formalistic, and costly. In actuality, they need not be so. Success is all in how to implement the regulations. Working smarter, not harder, applies to meeting regulations as much as to any other task.

Several cities, most notably San Francisco, have passed local ordinances concerning office ergonomics and work with visual display terminals (VDTs), that is, computers. The San Francisco ordinance was vacated in court with the premise that the VDT issues would be covered in the proposed California regulations. These VDT regulations are different from the proposed federal general industry standards in that they specify design criteria—heights of computer screens, features of chairs, and so forth.

American National Standards Institute

The American National Standards Institute (ANSI) is a private organization that oversees the development of industry consensus-making groups. For years ANSI has helped standardize countless items, such as the dimensions of components for consumer products so that the parts manufactured by a variety of companies would be interchangeable. ANSI has also helped set standards for safety, such as strength requirements for hard hats. Many standards first adopted by OSHA when that agency was founded in the 1970s were originally ANSI standards.

In the past few years, ANSI committees have been formed to set standards for ergonomics-related issues. A brief description of these activities follows.

> ### If Businesses Themselves Were Required
> Think what would happen if setting up a new business were reduced to a regulation. The requirements themselves would probably appear overwhelming and more costly than any possible gains. With limited resources, you would be required to:
>
> - set up a board of directors and meet regularly
> - write a business plan
> - plan and implement a marketing strategy
> - make financial projections
> - receive training in countless areas (for example, marketing, accounting, human resources)
> - implement the plan by producing the product or service (acquire space and equipment, hire personnel, find vendors for services and materials you need, and so forth)
> - evaluate progress
> - modify your plan
>
> These steps would appear intimidating if written out in full detail (especially if written by a bureaucrat who had never actually started a business). Yet, these "requirements" are what virtually all business advisers recommend as the means to success. Business magazines are filled with articles on taking these very steps.
>
> It is also true that there would be some people who would get so caught up in meeting the letter of the law that they would ignore the ultimate goal and fail to make any money. Success is found in the creativity by which the plan is executed.
>
> The above steps are simply the requirements of the OSHA meatpacking guidelines (and the assumed federal standards for general industry), translated into a business setting. When written down in a formal sense, they can appear to be worse than they actually are. But it is all in how you implement your plan.

Cumulative Trauma—ANSI Z-365

In 1991, ANSI accredited a committee to develop a standard for the prevention of cumulative trauma. Our work is currently in progress, but we expect impact in several areas:

- standardizing approaches to quantifying risk factors; for example, defining precisely what a "repetition" is so that everyone measures work tasks similarly
- determining the degree of consensus for the various aspects of CTD prevention programs; for example, formally agreeing on the factors that contribute to CTDs, or agreeing on the best approaches for evaluating and treating CTDs
- starting the process of setting specific numeric limits for the cumulative trauma risk factors, that is, how many repetitions, at what levels of exertion, in what posture, and so forth

Ultimately, the activities of this group are likely to provide the foundation for federal OSHA regulations.

Other ANSI Ergonomics Activities

ANSI committees have also been formed to address specific applications. These are:

1. *Visual Display Terminal Workstations (ANSI/HFTS 100–1988)*—ANSI and the Human Factors and Ergonomics Society (the professional association of ergonomists) cooperated to develop a standard for computer workstations. Much of the standard is highly detailed, defining standards for items such as screen and keyboard design that can be addressed only by manufacturers of computer equipment. Other parts of the standard, however, provide background helpful in understanding characteristics of good chairs and good dimensions for computer workstations.

2. *Ergonomics Guidelines for the Design, Installation and Use of Machine Tools*— This is a technical document, rather than a standard, and is in progress at the time of this writing. It provides guidance for the design of issues such as heights, reaches, exertion forces, clearance, layout, and location of controls and displays. The appendices provide additional guidelines for issues such as lifting, push–pull forces, and anthropometry.

It is possible that other committees will be formed in the future to concentrate on other, similar specific applications of ergonomics.

Americans with Disabilities Act

This act contains a number of requirements of employers to enable physically challenged people to work more easily. Most of these requirements affect hiring practices, but some concern ergonomics. In particular, the law obliges employers to make reasonable accommodations for individuals with special limitations.

Specific interpretations of legal questions have yet to be decided. However, two things are clear:

1. Employers need to know more about the requirements of the different tasks in their workplaces, and

2. Employers must be able to make "reasonable accommodations" for disabled employees.

Regardless of eventual court decisions, the general direction is clear and the value of ergonomics is affirmed. The tools of ergonomics can be used both to (1) characterize job requirements, and (2) help conceptualize possible improvements to permit disabled employees to contribute to the economy.

Consistent with the argument of the rest of this book, the goal of accommodating individuals with disabilities can be met in a way that makes good business sense, if (1) true costs and benefits are understood, and (2) common sense and creativity are applied.

Litigation

Finally, another strategic trend of concern for management is the increasing number of lawsuits related to ergonomics issues. Product liability suits have been filed alleging that products have been "defective" if they are not "ergonomic." Furthermore, some plaintiffs have argued that the absence of good ergonomic design in the workplace has constituted negligence.

Conclusion: Present Management Needs

Regardless of the merits of any details of the above issues, the trends are unmistakable. Managers need to use this window of opportunity and educate themselves about ergonomics and how it can help them address the specific issues in their own operations. Ergonomics can help managers realign their strategies to meet changing human resource challenges and compete successfully in the global economy. Innovative thinking can enable managers to implement regulations in a way that makes good business sense.

Schapter
SEVEN

AN ESSAY ON WORK
ORGANIZATION

This chapter provides a number of insights on management
in general based on an ergonomic perspective. It introduces
the topic of work organization, then focuses on specific
points. In particular, it suggests that the phenomenon of
mental stress provides a good framework for thinking about
work organization. The chapter concludes by showing that
the solutions parallel the Total Quality Management
methods.

WORK ORGANIZATION
IS A HUMAN TOOL

Just as humans have developed physical tools and technology throughout history, they have also developed successive ways to *organize* in order to expand human capabilities and overcome human limitations. In this sense, organization itself is a human tool, and thus a good focus for ergonomics.

Work organization can be thought of in an anthropological sense— the various ways humans have devised to perform tasks and obtain the necessities of life. We might think, for example, of the division of labor among hunter-gatherers, then how people reorganized this division of labor with agriculture. We can think of the differences in how work was organized during the early stages of the industrial age versus the present.

Furthermore, we might also think of all of the different ways that various cultures around the world have organized their activities at each stage of this development—different societies have made different choices about how work was to be accomplished. One can, for example, compare the ways in which industrial work has been organized in the former Soviet Union versus those in the United States.

Work organization thus refers to the underlying design of work, encompassing a range from job-specific issues to transcending systems:

1. Microdecisions, such as task allocation, that are made day to day by millions of people.
2. Administrative practices, such as compensation plans, that are planned as part of a business strategy.
3. Overall management systems that are often taken for granted as part of the philosophical basis of modern industry.

Examples of issues include:

- *Task allocation*—How should tasks be divided and assigned to accomplish goals? Is it better to have many people equally capable of doing many tasks? Or is it better to have a narrow division of labor, so that individuals can be extremely highly qualified at specific tasks?
- *Assembly line or work cells*—Should the technology and equipment of the workplace be designed so that tasks are narrowly defined? Or should the physical layout promote team activities?
- *Shift work*—Should there be more than one shift in a given workplace? And, if so, should employees be assigned to just one shift (thus prohibiting some people from enjoying normal evening family and social activities)? Or should they be rotated between shifts every couple of weeks (thus requiring everyone to disrupt their biological time clocks)?

- *Reward system*—How should people be compensated for their activities? What are the actions that are rewarded? Should people be compensated for how much they put into a task (hours and effort), or how much they put out (quality and quantity of product)?
- *Structure*—How many vertical layers should there be in an organization? What degree of horizontal segmentation? What amount of centralization?
- *Decision-making*—What kinds of decisions should be made at what levels of the organization? Should the strategic issues be left to just top managers? Or should rank-and-file employees be allowed—or required—to take part in decision-making?

These are the kinds of organizational issues that are encompassed by the term *work organization*. Again, we can think of each method of approaching an organizational issue as a tool.

What Ergonomics Has to Do with It

As with any other tool, ergonomics seeks to analyze and improve the fit between the system of organization and the people who make up that organization. In parallel with purely physical issues, the goal here is to design an organizational system that matches human requirements and thereby increases (1) overall efficiency of the system and (2) personal fulfillment and well-being of people.

Certainly other fields of study have addressed these issues, but ergonomics provides a special perspective. Furthermore, as outlined in Chapter Two, much of how humans have developed organizations over the past 40,000 years has been instinctive—seeking the best way to organize human activity for efficient production, generation of wealth, and procurement of food, clothing, and shelter.

But not everything is instinctive, and specific organizational structures are not always human-friendly. What lies before us is to use the perspective of ergonomics, with its expanding set of analytic tools, to address aspects of modern production systems.

More than other fields of study, ergonomics focuses on the point where both the technological side of production and the human side intermesh. For example, the fields of management science and organizational psychology address many of the people areas, but do not always capture the technical side. Conversely, engineering addresses technical factors, but does not always capture the human side. The combination of the two is what interests ergonomists—the interrelationships of humans and their tools.

Organization as a Flexible Tool

Technology structures what is possible, but it certainly does not predetermine every aspect of an organizational system. We have seen in recent years that the most efficient method of production of products such as automobiles is not always that of an assembly line. Just because assembly line technology exists does not mean that it should always be used, or that it is even the best way to assemble products.

Other methods of organizing work are possible. We take much for granted—the assumptions of twentieth century America—because that is the way things have been done for 100 years. Nonetheless, there are still many ways to organize modern industrial work. The value of thinking of organization as a tool is that it puts our organizational systems in the same league as physical items such as software and furniture. In so doing, work organization becomes a factor that can be modified and made more friendly to benefit both efficiency and human well-being.

A truly innovative company or society holds these organizational issues as fair game for change. And ergonomics can provide the same challenges to assumptions about the organization of work in order to promote innovation as it has to promote the development of software and furniture.

Technology is changing at a more rapid pace than ever before. There is a crucial need to evaluate the options for organizational structures to take best advantage of the new technology. Addressing work organization opens more doors and raises more opportunities. Choices are being made, and ergonomics can help identify those that are both more efficient and most compatible with human fulfillment and well-being.

Taylorism—The Separation of Planning from Doing

The framework for work organization that has pervaded modern times has been known as Taylorism, after Frederick Taylor, an engineer at the turn of the

---

TERMINOLOGY

Ergonomists have used other terms besides *work organization* to describe this topic:

- *Sociotechnical systems* has been used in particular to describe the intertwined area of organizational structures and technological development.
- More recently, *macroergonomics* has been used for this whole area. (The prefix *macro* is used to contrast the broad issues with, for example, the microdetails involved in the design of computer keyboards.)
- Some researchers use the term *psychosocial* to refer to the human aspects of the workplace, as opposed to the strictly physical and technical side.

All in all, these terms mean much the same thing. *Work organization* is the term I use to mean all of the above.

century. Taylor, the inventor of time and motion study, developed an entire system of management based on this analytic technique.

Taylor's book, *The Principles of Scientific Management,* published in 1911, is required reading for anyone interested in these issues. The core of the book describes his efforts to analyze the work methods of employees and, through careful structuring of each step of a task, teach the employees the most efficient technique considered possible. Widespread use of this time-method analysis contributed to rapidly increasing rates of productivity in the first part of the twentieth century.

There is much of great value in Taylor's concepts, including many aspects of his system that have been misrepresented. For example, in his book Taylor made a strong case that employees should benefit equally with owners in sharing the rewards of increased efficiency. However, in practice that has not always been the case.

But many of Taylor's thoughts on management now seem out of date and have been challenged in the past few decades. The essence of Taylorism has been the dominating role of management in determining the best way to perform a task.

When Workmen Ran the Shops

Prior to Taylor, work was organized quite differently than it is today. In his book, Taylor described "unscientific" worklife in the steel industry in the late nineteenth century. He penned many statements that come as a surprise today, including, "As is still usual in most of the shops in this country, the shop was really run by the workmen, and not the bosses."

Managers' jobs in that era were to get orders; the workers were left to decide how best to produce the goods based on their own hand-me-down training. Taylor showed that efficiencies could be vastly improved through his analytic techniques. Subsequently, one of the tenets of Taylorism was to separate planning from doing. It became part of management's function to plan the steps of a task down to the most minute details.

Although these efforts were highly successful in the first part of the century, in the latter part of the century they have led to inefficiencies. Taylor believed that managers were needed to do the planning simply because employees did not have the education to use his analytic tools. Today people are better educated and can easily be taught analytic and planning tools, especially for those issues related to their immediate jobs. Continuing to separate planning from doing has deprived industry of a considerable source of innovation and has ultimately created unneeded layers of management.

One Hundred Years to Overcome

Today, 100 years after Taylor, modern work organization is clearly moving toward reintegrating planning with doing. Increasingly, employees are becoming involved in decision-making. They are being asked to help analyze and plan, instead of just being told what to do. These efforts have been known in the United States variously as quality of work life (QWL), quality circles, empowerment, autonomous work groups, and the like. Despite fits and starts and the failure of some specific programs, the trend is clear. It has taken 100 years to develop the current system and it will take more than a few decades to shift to a new framework.

Time and motion study still has a role as an analytic technique in industry, as one of many tools such as statistical process control or root-cause analysis. Motion study is clearly a part of the effort needed to reduce CTDs. Even time study can be useful despite criticisms of the traditional method, as long as there is sufficient emphasis on employee well-being. We now understand that certain postures and repetitive motions can be injurious, which was not taken into account in traditional time and motion study. Taylor did state that, "in no case is the workman called upon to work at a pace which would be injurious to his health." But cumulative trauma was not recognized in his day. Time and motion standards probably do not give sufficient rest and recovery time for highly repetitive tasks and there is a need to reevaluate systems based on new information. However, the tools still have value in the proper context.

As a final comment, Taylor has been misinterpreted in one other way. People have assumed that his recommendations to have managers do the detailed planning of job tasks also meant that he believed workers should not provide any input. On the contrary, Taylor encouraged employees to make suggestions for improvement and he believed in providing bonuses for ingenuity.

Taylor compared his system to that of training surgeons. Surgeons are taught procedures down to the most minute details, Taylor wrote, but are then able to use "originality and ingenuity to make real additions to the world's knowledge, instead of reinventing things which are old." He implied that through his system, factory workers could ultimately do the equivalent. Undoubtedly, as the system was applied, many people got the impression that workers were supposed to stop thinking, but Taylor himself had a different view.

Workplace Stress—A Criterion to Evaluate Work Systems

Work organization is a broad, generic term that includes controversies, studies, and a tremendous amount of opinion. There are many ways in which work can be evaluated. Some approaches are based on objective criteria, such as studies of production output based on different methods of organization. Similarly, many aspects of work organization can be evaluated based on ways to improve product quality.

At times, issues are decided without any objective basis at all. Many management practices have been established simply based on opinion. Too often systems have been based on the whims and egos of individual owners and managers. A notorious example is the tendency of some managers to establish authoritarian systems based on a need to control others, whether the system is efficient or not. Expediency at the time represents another manner in which issues are decided—systems are established without much forethought of any sort.

Within such a huge area, there is more information than a single book can hope to address. However, it is possible to suggest one criterion by which many issues of work organization can be evaluated.

The ergonomics contribution is that the framework should be based on human requirements. Indeed, the studies of work organization that do not take the human response into account may be inherently flawed.

A number of approaches are possible in this regard, but the one I offer here is that of mental stress. As we will see, the focus on mental stress establishes a perspective on organizations based on human physiology. Then, through a series of steps, the study of stress can lead to a conclusion showing the value of the framework of Total Quality Management in designing out stressful situations. To start this sequence, we begin with the physiological stress response.

WORK ORGANIZATION AND MENTAL STRESS

Mental stress, properly defined, provides a useful perspective for reviewing much of how humans fit into the workplace. It serves as a common denominator for addressing a multitude of work organization issues. This section provides an explanation.

The Stress Reaction

The popular usage of the term *stress* has lost much of its original scientific connotation. As an unfortunate result, the word has become a vague catchall for a jumble of anxieties and pressures. Consequently, the concept of stress has been overused and trivialized, creating some legitimate skepticism about the topic.

However, the phenomenon is significant in human behavior and has great relevance for the design of effective organizations. It is important to grasp the precise meaning.

Stress is best understood as a physiological reaction. When confronted with a threat, the body responds with a variety of physiological changes, including increases in adrenaline, heart rate, blood pressure, respiration, and muscular tension. In essence, the body gears itself for a physical attack based on a perception of danger. Keeping this physiological reaction in mind is the key to understanding "mental" stress.

This stress response can be beneficial, such as by preparing the body to protect itself from a threat. Sometimes known as the "fight or flight" syndrome, this reaction is fairly standardized in all animals as a defense mechanism. In an evolutionary sense, this mechanism has been quite adaptive. The individuals with the strongest stress reactions were better able to defend themselves against physical dangers. These individuals survived and passed along these traits to their offspring. Consequently, all of us in modern society have inherited this biological phenomenon.

Unfortunately the stress reaction can be triggered by events other than a physical threat from which we need to flee or fight. The stress reaction can occur in everyday life in response to a wide variety of stimuli, including daily verbal battles, unsettling television news, and continued pressure for personal performance. At times, this reaction can still be helpful, such as providing the body with enough adrenaline and energy to finish a task by a deadline. More typically, however, the stress response is not helpful to the individual—the body's reaction to prepare to flee or fight is irrelevant to the task at hand.

Stress and Disease

If the stress reaction is prolonged in an individual and the physiological changes in the body continue unabated, a number of harmful effects can occur. These consequences are primarily illness, but also include maladaptive behaviors such as absenteeism or overeating. The links between prolonged stress and these illnesses have been well studied. Table 7–1 lists a number of the specific physiological changes that constitute the stress response as well as the parallel disorders that can occur when people are under chronic stress.

Gaining an understanding that the stress response affects the entire body helps explain why the consequences of stress can be manifested in so many different ways. Keeping this in mind helps prevents the issue from degenerating into pop psychology. Mental stress is a physiological reaction.

Common Sources of Workplace Stress

Over the past few decades considerable research has addressed the sources of occupational stress. The list of these stressors is long, but has great relevancy for developing good management systems.

Time pressures and work overload constitute the stereotyped sources of job stress. Excessive work loads, either too much or work that is too difficult, are clearly linked with the commonly recognized stress disorders. However, there is much more. Work underload is less obvious, but surprisingly is equally important as a source of stress. Boredom and lack of opportunity to use skills can lead to physical symptoms of illness. Machine-paced work (as opposed to self-paced work), epitomized by the assembly line, has been identified in a variety of major studies as highly stressful. Some of the most stressful jobs are those that require high levels of

TABLE 7.1   *Physiological Changes and Associated Diseases*

Physiological Change	Associated Diseases
Increased heart rate and constriction of blood vessels	Cardiovascular disease
Increased acid and enzyme secretions in the gastrointestinal system	Peptic ulcers
Muscle tension	Back and neck pain
Changes in hormonal secretions	Infectious and degenerative diseases
Constriction of cranial blood vessels	Migraine headaches
Drying of mucous membranes	Allergic reactions
Constriction of peripheral blood vessels and changes in the electrical conductivity of skin surfaces	Dermatitis

mental concentration to perform elementary and highly repetitive tasks, paced by machine. Additional stressful situations include excessive overtime, wage incentive systems, and extremely close surveillance of employees, such as electronic monitoring systems.

The physical work environment can also create stress. Distracting or constant high levels of noise in particular can be perceived by the body as a warning and can trigger the stress response. Lack of privacy and crowding of people are similarly related.

Interpersonal relationships are also common sources of stress. Interactions with supervisors, co-workers, and subordinates can be challenging or threatening and therefore stressful. Closely related are anxieties over one's role in an organization,

THE COSTS OF STRESS

Stress is a growing problem. A study of 1299 employees concluded that:

- Four of ten American workers say their jobs are "very" or "extremely" stressful.
- Fifty percent of employees believe that job stress reduces their productivity.
- Those who report high stress are three times more likely to suffer from frequent illness than workers reporting low stress.

Northwestern National Life (1992)

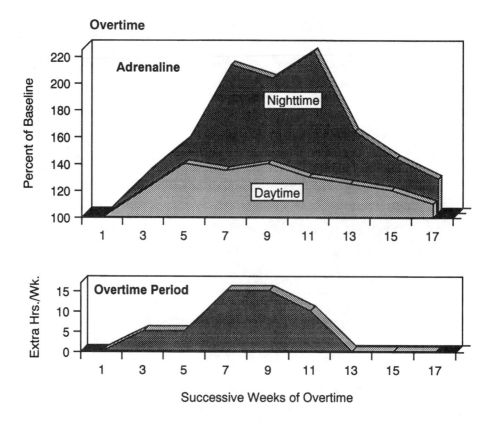

This graph shows the pattern of adrenaline levels from employees who worked a period of overtime. After the overtime was initiated, the employees' adrenaline rose. Subsequently, as the overtime peaked and then decreased, the levels of adrenaline followed suit, although with a bit of a lag time. (*Source:* Frankenhäuser, 1981.)

especially if there is "role ambiguity" (not knowing what one's role is) or "role conflict" (differing views by others of one's role). Working in larger organizations seems to be more stressful than working in smaller ones, perhaps because of an increased likelihood for role ambiguities and conflicts as well as a reduced sense of belonging. Job insecurities and even changes in jobs have been linked with stress. Organizational disruptions, such as frequent changes in management or labor–management disputes, can easily be understood as stressful.

Results of several studies are shown in the accompanying graphs. These help depict the scientific basis for understanding occupational stress.

Our bodies can perceive all of these factors as threatening in some way or another and react, even though that reaction is often not particularly helpful to the situation at hand. Our bodies cannot distinguish between a saber-toothed tiger and the news of an impending plant shutdown. All we perceive is danger and we respond, ready for "fight or flight."

A common thread through many of these stressful situations is the individual's

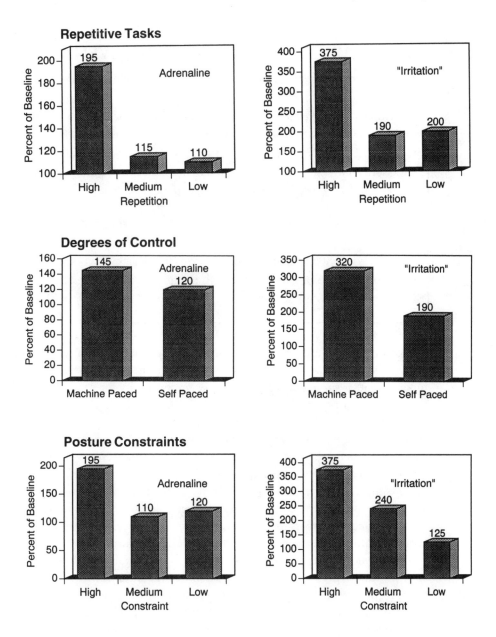

There are two common approaches to measuring workplace stress: (1) physiological measurements such as adrenaline excretion and (2) employee survey questionnaires. The above graphs show results from both approaches—the levels of adrenaline from employees working at various tasks as well as subjective reports of feelings of irritation. Note that the objective measurements of adrenaline generally match the subjective statements of feelings. In particular, the above graphs show the effects of various work organization factors on stress levels. Data of this sort provide support for the concept that tasks that are highly repetitive, machine-paced, and require employees to work in constrained postures are stressful. (*Source:* Frankenhäuser, 1981.)

ABILITY TO AFFECT DAILY EVENTS

A particularly interesting study highlights the underlying importance of being able to control events to the extent possible. In this study, a number of individuals were exposed to high levels of noise. Some of the people were provided with a switch that could be used to stop the noise, while others were not. The study showed that those individuals with a switch experienced less stress and performed better than those without a switch, *even though the people actually never used the switches.*

Glass and Singer (1972)

This concept has many implications for the design of work. If we can find more ways to provide people with control over daily events in the workplace, stress can be reduced.

lack of control over many events of daily worklife, ranging from the pace at which one works to the effects of an economic recession. One of the most stress-provoking predicaments is simultaneously having the responsibility for a task, but not having the means to carry it out.

Although there is a common perception that managers are highly stressed, the evidence shows that production and office workers actually experience more stress. Again, the difference turns out to be the ability to control events. Thus, despite the fact that managers may experience more pressures and may be confronted with more complex problems, as a general rule managers are in a position to do more about the problems and consequently experience less of the stress response. In contrast, production workers may deal with more mundane issues, but have less ability to do anything about them and therefore experience more stress.

As an additional comment, a number of factors can buffer the stress response. An individual's personality and perception of events can sometimes affect the response. Some people are genuinely calm in a crisis, while others seem to overreact to every minor issue. Social support can also serve as a buffer. Talking to friends and work colleagues helps, as does getting their sympathy and understanding. Additionally, exercise and good health maintenance can alleviate the effects of stress.

Engineering Out Sources of Stress

The ergonomic perspective on workplace stress is to find ways to design systems to prevent the sources of stress. Again, we must recognize that sometimes stress is helpful, such as for getting projects out on time. But constant stress can be an ongoing drain on individuals, and ultimately the organization's bottom line. Stressful situations can, and should, be prevented.

The ergonomics approach contrasts sharply with stress management efforts in recent decades. The typical efforts over the past number of years have been to address what the *individual* can do to *cope* with stress, for example, with exercise and relaxation. Moreover, it is commonly left to the individual to initiate the stress reduction activity, with all of the motivational problems that entails.

But ergonomics is all about design, and the ergonomist focuses on how tasks can be designed to eliminate sources of stress. There are many options in this regard. Some purely physical equipment changes can be made, such as providing stop switches on production lines. More to the point here are improved systems of work organization, for example, by reducing overtime. Finally, we can build into the organizational system ways to buffer the effects of stress, perhaps by providing regular stretch and exercise breaks.

The task at hand, then, is to identify the sources of occupational stress and determine how best to engineer out these situations. It is tempting at first to merely

TABLE 7.2 *Positive Approaches to Work Organization That Overcome Stressors*

Stressor	Design Criterion
Time pressures and work overload	Good planning
Work underload	Job enlargement
Assembly line	Work cells
Lack of control	Employee involvement
Overtime	Good planning and proper staffing
Piecework	Reward process improvement
Close surveillance of employees	Employee empowerment
High noise	Improved equipment
Lack of privacy and crowding	Space planning
Interpersonal relationships	Team building
Role ambiguity and conflict	Good communications
Large organizations	Team-based work spaces
Stress Buffer	**Design Criterion**
Social support	Employee assistance programs
Exercise	Regular stretch and exercise breaks
Health maintenance	Wellness programs

list the negative of the problem—avoid overtime, avoid machine-paced work, and so forth—but that does not get us very far. It is more rewarding to try to define a positive approach to work organization that overcomes the problem, as show in Table 7–2.

The list of Table 7–2 can be expanded, and there is more than one way to respond to each issue. But it should be evident that it is possible to prevent stress through better work design. It is also true that some improvements are easier said than done, but at least we can move in the correct general direction and identify the goals.

The Quality Improvement Process

A more important conclusion can be drawn from Table 7–2. Virtually each of the organizational design criteria there has been introduced to industry as part of the quality improvement process or other modern approaches to running organizations. The list of Table 7–2 looks like the chapter headings from a recent textbook on Total Quality Management.

To be sure, some issues such as shiftwork fall outside this framework. Furthermore, issues such as machine-paced assembly-line-style work have not been emphasized as much from other perspectives. Nevertheless, the following general point still holds true—the efforts needed to prevent workplace stress are the same as those of other advanced approaches to human resource management.

Conclusion

The efforts needed to prevent occupational stress basically amount to good management, especially using the perspective of Total Quality Management and recognizing that these general parallels can be drawn between quality and ergonomics.

In some ways there is nothing particularly new here, but it is reassuring that guidelines for organizing work based on human physiology are generally the same as those that have been developed based on other criteria, such as quality. Business and industry can proceed with introducing and improving these quality-based management initiatives with even greater confidence than before.

STRESS AND CUMULATIVE TRAUMA

Scientific research provides support for adding stress to the list of risk factors for CTDs. Although there is much investigation needed to identify the precise relationship between these issues and CTDs, there appear to be certain stressful situations that increase the risk for cumulative trauma.

A variety of studies have been conducted that link CTDs with psychological

---

JAILHOUSE SECRETARIES

One time I was asked to review CTD problems in the city hall of a major metropolitan area. As we reviewed the records for wrist disorders, we discovered that most of these cases came from the police department. In turn, we found that most of these cases related to one specific job.

The job in question was that of typing information provided by police officers as they brought in suspects to be jailed. A number of psychological stressors were present.

First, criminals tend not to be arrested on a scheduled basis, so for periods of time during the day, the secretaries on this job had little or nothing to do ("underload"). Then suddenly, the officers would bring in a group of perpetrators whom they had just arrested. All at once there was a need for the secretaries to type fast and finish their tasks ("overload").

Moreover, the room was small and the secretaries were only slightly beyond striking distance of the often abusive suspects ("physical threat"). You could almost see the tension in the secretaries' hands and wrists as they worked.

Thus, these secretaries had the factor of significant stress in addition to all of the usual physical issues involving chairs, desks, and keyboards. There was really no mystery why they might have been experiencing greater problems than other secretaries.

The recommendations for improvement were simply for officers to use tape recorders to dictate their reports and to change the layout of the room to provide a greater physical barrier between the secretaries and the criminal suspects. These ideas were in addition to the usual office ergonomic improvements.

---

stress and work organization issues. It appears that stressful situations like overtime, incentive pay, and close monitoring of employees can contribute to the development of CTDs. Stress alone does not seem to cause CTD problems. Its effect is to amplify the severity of the physical risk factors.

Although there are scientific issues and tests that still need to be resolved, the relationship between stress and CTDs appears solid. Consensus on this topic is growing as evidenced by the American National Standards Institute (ANSI) Z-365 Committee on Cumulative Trauma, which has accepted stress as a risk factor.

The mechanism of action appears to be that when the human body is under stress, certain muscles may be constantly tensed, which in turn intensifies other risk factors such as forceful or repetitive motions and creates higher rates of CTDs.

SIGN LANGUAGE INTERPRETERS

There was a hearing-impaired worker and a sign language interpreter among the 50 or so participants at one of my presentations in a large factory. After the session, the two came to me and asked if I was aware that sign language interpreters had high rates of wrist problems. I was not, and the interpreter proceeded to explain to me that of her group of eight interpreters, six were having problems.

I took this in, nodding my head in understanding. Then, however, they asked the question that was really on their minds: "Why is it that interpreters are having problems, but deaf people themselves do not seem to be affected?" I was puzzled, then answered that perhaps the interpreters used more exaggerated wrist motions in order to be seen by a larger group.

Later, in conducting a literature survey (see "Studies on Work Organization and CTDs"), I found an additional piece of evidence. Sign language interpreters work in an inherently stressful situation—conducting simultaneous translation, often on topics with which they might not be personally familiar. I reflected back on my own presentation and the specialized terms I used and wondered how she signed these.

This added stress of simultaneously listening and translating may well create a situation quite different from that experienced by the hearing-impaired. It is also different from that of deaf instructors, who may need to make exaggerated motions to be seen by students, but do not need to simultaneously listen and translate.

Stressful situations can, in fact, increase the risk of cumulative trauma by about two or three times, given equal exposure to physical risk factors. (See "Studies on Work Organization and CTDs" and "Stress in Nonindustrial Societies" at the end of this chapter for a full literature review on the above topic.)

Thus, preventing CTDs involves more than just physical issues like tools and workstation layout. Work organization issues must be addressed as well.

The added risk factor of stress explains many mysteries. Many investigators have noted over the past decade that the physical risk factors for CTDs (for example, force, repetition, awkward postures) do not always explain the differences in rates of CTDs between different work activities. That is, sometimes tasks are equivalent in terms of physical activities, but have differences in CTD experience. For example, these tasks might be two identical work processes located in two different facilities in two different parts of the country. Or they might be very equivalent jobs side by side in the same workplace. Yet people in one situation seem to have more CTDs than the other.

These variations can be explained in several ways:

- random differences in individuals
- differences in the levels of awareness among groups of people
- differences in the willingness to report problems by some people or in different parts of the country

However, sometimes these puzzles are better explained by the presence of an additional factor—the various aspects of work organization that appear to be involved. The underlying factor is any situation that increases psychological stress levels among employees.

Although it is heartening that some puzzles are explained by recognition of this added factor, it does mean that more issues than simply engineering changes need to be addressed in a workplace CTD prevention program. Work organization issues must be changed as well, some that cut to the heart of management philosophies.

Fortunately, as we have seen in the previous section, the kinds of organizational changes needed here are the same as those being encouraged as part of Total Quality Management.

WORK ORGANIZATION AND PREVENTING CUMULATIVE TRAUMA

The following section expands on a number of work organization issues that commonly need to be addressed in a workplace effort to prevent cumulative trauma.

There are three ways that work organization factors can increase the risks for CTDs:

1. *Increased Physical Risk Factors*—Certain work organization factors can increase the exposure to physical and environmental risk factors. The best example is overtime, which simply increases the duration of exposure and, therefore, affects the number of repetitions within a work period and reduces the available recovery time. Wage incentives may also increase the number of repetitions without changing the duration of exposure. Longer hours and wage incentives can also modify the work/rest cycles.

2. *Employee Behavior Changes*—Certain organizational factors can affect the way that people perform work. The concern is that these behaviors result in increased exposure to the physical risk factors. For example, it is not uncommon to see workers perform shortcuts to increase incentive earnings that raise the risk of injury. Machine pacing can induce workers to "work ahead" constantly reaching for parts. A worker who is new on a job will often use inefficient and unnecessary movements, or may overtighten, or overinspect items. Inspection workers and keyboard operators adopt more rigid postures with increased pace or speed. Operators will also strike controls or keys harder than necessary.

3. *Mental Stress*—The third way that work organization factors can affect the risk of CTDs is by increasing the level of psychological stress, as described in the previous section. Once again, the issue here is not that stress can itself cause a CTD, but that it can exacerbate the effects of the other physical risk factors.

To be clear on this matter, the core of a workplace ergonomics program to prevent CTDs should still concentrate on physical engineering factors, such as changing tools and equipment to reduce repetition or enable employees to work in good postures. In some workplaces, however, there may be special work organization issues that should also be addressed. Several issues are briefly described below, some of which are common problems and the others are standard solutions often raised with regard to CTD prevention programs.

Common Problems

Excessive Overtime

One common problem is excessive overtime for highly repetitive or strenuous jobs. As we have seen above, overtime contributes to CTDs in three ways. Overtime hours can (1) increase the daily dose of repetitive motions, (2) reduce the time available for body tissue to recover, and (3) trigger the physiological stress response. Despite short-term cost savings by increasing overtime rather than hiring additional staff, there can be long-term consequences and perhaps ultimately higher costs.

The problem here is not temporary periods of overtime, such as peak seasonal production. Rather, the difficulty lies with decisions by managers to require overtime on a routine basis rather than hire additional staff. Exceeding a normal workweek may surpass people's biological abilities to recuperate and replenish themselves under some circumstances.

Electronic Monitoring of Employees

Extremely close surveillance of employees has also been linked directly to increased levels of stress, which in turn increases rates of CTDs as well as other problems. Two examples of controversial practices are:

* supervisors listening in on employees' telephone conversations with customers
* electronic monitoring systems—software that keeps track of keystrokes for clerical employees

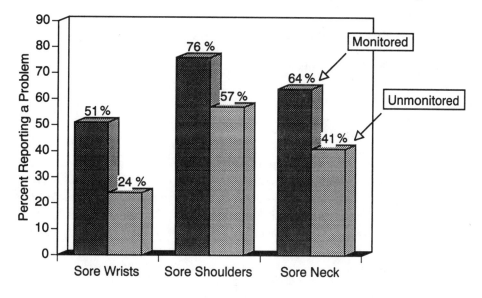

Results showing that employees who work under close supervision report increased symptoms of cumulative trauma than do employees without such supervision, despite having the same types of jobs. (*Source:* Smith et al., 1990.)

Such practices do seem out of place in the context of Total Quality Management. A more typical trend in industry is to empower and involve employees to increase self-motivation. In a positive environment like this, practices such as electronic monitoring of employees may be irrelevant.

Wage Incentive Systems

Wage incentive and other compensation systems need review to ensure that employees are rewarded for the right behaviors. A particular issue is the traditional piece-rate system, which has been detrimental to both product quality and employee well-being. The reward system should encourage employees to focus efforts and skills on a broader range of activities other than just producing pieces as fast as possible.

It is important to emphasize that incentives and rewards per se are not the issue. Rather, the issue is determining which behaviors to reward. In most quality-oriented management systems, for example, incentives are designed to reward continuous improvement of the production process, rather than solely production quantities.

Common Recommendations

A number of work organization issues are often raised as part of the CTD prevention effort.

Job Enlargement

One of the reasons why repetitive trauma has increased in recent years is that the subdivision of labor has often become overemphasized. In many cases, it would be better to have employees perform more parts of a job than merely one specific task repeatedly, that is, we should "enlarge" jobs rather than make them increasingly narrow. Job enlargement has a variety of additional benefits, and is a strategy pursued by many companies throughout industry. The classic example is the switch from assembly-line-style production to a team or cell concept.

Job Rotation

Job rotation may also help in come circumstances. There are many advantages to rotating workers, including cross-training, increased job stimulus, and better un-

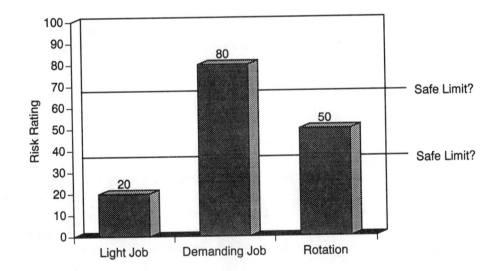

Graph depicting the "averaging" effect of job rotation. Note that the resulting rotation job, which is the average of the light and demanding job, may be safe or it may be dangerous depending on where the actual safe limit is. At this point, science does not know this answer.

derstanding of the organization's whole operations. Job rotation can work well to prevent CTDs especially if there are only a few demanding tasks among a group of light ones.

However, there are a number of concerns with regard to this concept. From the standpoint of preventing CTDs, job rotation is only beneficial if the tasks involve different muscle–tendon groups or if the workers are rotated to a rest cycle. Rotating from one wrist-intensive task to another, for example, may provide very little or no relief.

Poorly structured job rotation programs, may, in fact, *increase* the risk of CTDs. If employees are not properly trained or accustomed to the tasks they are to do, they can increase their exposure to the risk factors (see "Break-in Periods," below). Switching to new tasks to which employees are not prepared, or not rotating frequently enough to stay accustomed to the task, can thus increase the physical stress of the job.

Furthermore, job rotation alone does not change the risk factors present in a facility. It only distributes the risk factors more evenly across a larger group of people. Thus, the risk for some individuals can be reduced, while the risk for others is increased. There will be no net change in risk factors present.

When employees rotate between two jobs the risk exposure can be thought of as being "averaged." Job rotation may drop the average to within a safe level, or raise the whole group in excess of safe limits. Unfortunately, it is not possible with current knowledge to determine what the safe limit is. For this reason it is critical to select job rotations that minimize the exposure level.

Setting up a good job rotation system is thus not a trivial project and should not be undertaken lightly. Jobs must be evaluated carefully, and employees properly trained, involved, and allowed a sufficient break-in period. Finally, although job rotation may have beneficial effects, engineering changes should remain the goal of the ergonomics program. It is vital to continue efforts at changing the physical makeup of the jobs.

U.S. UNION WORK RULES

Many U.S. unions have been critical of job rotation programs. Some of their concerns are identified above—inadequate selection of jobs for rotation schedules, inadequate training and break-in periods for workers, and finally, job rotation that really does not get at the source of the problem.

Furthermore, there is a common statement of many high-seniority people when they are asked to rotate that is easy to empathize with: "I worked years on those harder jobs and finally earned the right to an easier job. Now you're asking me to go back to that harder job again? I've paid my dues."

These issues can, and should, be addressed in any organization that is considering establishing a formal program.

But unions also base some of their opposition to job rotation on the belief that the traditional U.S. practices regarding job classifications are somehow based on union principles. There are many ironies in this belief, since when we look at the historical roots of these practices, we actually find what was once a management approach to improving productivity. What we in the United States now think of as union work rules are a management invention, not a union one.

The current system of job classification flows from time study and other efforts to fragment work into its simplest possible forms. As we have seen previously, in the nineteenth century jobs were much broader and more flexible than they are today. With the rise of Taylorism and other trends, job classifications became much narrower.

World War II intensified this trend, when millions of inexperienced people were brought into industry. Jobs were further simplified and subdivided to enable the work to get done. Subsequent rulings by the National Labor Relations Board (NLRB) in the postwar era brought additional emphasis to these job classifications and institutionalized their form.

Simultaneously, unionized workers realized that these new systems inadvertently offered some protection against overzealous and arbitrary management attempts to speed up work and reduce the work force. Reliance on these job classifications also protected workers from being assigned to jobs for which they had insufficient training, in particular, jobs at which products could easily be damaged (and thus provoke dismissal) or jobs that were potentially dangerous.

As the generations passed, both unions and management collectively forgot the origins of this system. Now both parties view the framework for job classifications as a union, rather than management, program.

In contrast, many European trade unionists find the U.S. system unusual. In particular, in Scandinavia (with some of the strongest labor movements in the world) there is no equivalent to the U.S. practice. These other labor movements have through collective bargaining and legislation found ways to work with management to provide the needed protection of individual workers, yet maintain shop floor flexibility.

Break-in Periods

Unaccustomed activity is an issue that concerns many workplaces. Workers new to certain tasks need time to build up endurance and learn the nuances that make tasks easier. If no break-in period is provided, certain types of CTDs can occur, such as tendinitis.

For jobs that require a high level of skill, this break-in period occurs naturally, since the employees need a slower introductory period anyway to learn to complete the task at full speed. The problem is with simple and highly repetitive jobs, where employees can physically perform the work at high speed within a few minutes or hours, but have not had time to learn subtle techniques of the task or become accustomed to using certain muscles. Even people who have previously done a task, but not recently, can be affected when they are reassigned to the task.

The length and nature of a break-in period depends on the requirements of the specific task and the capabilities and experience of the individual employee. In time, muscles become conditioned, plus develop "muscle memory" to perform the task. In addition, people learn the tricks that reduce the exertion and motions needed.

INNOVATIVE BREAKS IN A SLAUGHTERHOUSE

An example of an innovative approach to providing short breaks comes from the meatpacking industry. On the slaughter floor, all of the jobs are in sequence, following a chain on which the carcasses hang. The minibreaks in one facility work in the following fashion:

The first job in the long sequence is the stunner, who stuns the animals prior to being butchered. The stunner begins the minibreak process by stopping for a few minutes every hour, then starting up again. This leaves a small gap in the chain that proceeds to work its way through the department. As the gap reaches individual employees, they can stop for those few minutes, stand back and stretch, then begin working again.

Since the employees do not have time to leave the area, and only a few people at any one time have stopped, these breaks are quite easy to administer and the usual traffic jam to the canteen is avoided. (These short breaks are in addition to their normal breaks.)

Work/Rest Schedules

Another issue is how breaks are provided, both in terms of the number of breaks and the length of each. Taking several short breaks is better than a single long one, both for reducing fatigue and cumulative trauma and for improving productivity. Muscles and tissue get a chance to recover and mental capacity is refreshed. Even very short breaks—"microbreaks"—may have a positive effect.

Working too long before taking a break can cause excessive fatigue, from which even a long break cannot provide refreshment. Almost everyone who has attended a long training conference can attest to the fact that productivity drops when there are insufficient rest breaks.

Precise information on optimal work/rest schedules is not known in a way that can be generalized to all tasks and industries. The only rule of thumb is that many short breaks are better than a single long break.

Exercise

Many companies are beginning to provide exercise during the course of the day:

- *Warm-ups*—to limber up muscles before performing strenuous work, much as an athlete world do.
- *Regular stretch breaks*—to periodically stretch unused muscles and provide relief from being in the same position for too long.
- *Conditioning*—to get people in shape by strengthening muscles and improving flexibility and endurance.

There are a variety of reasons to provide for such exercise, including overall employee wellness and morale, both of which can result in cost savings from lower health care premiums, lower absenteeism, and increased productivity. Additionally, well-designed exercise programs can serve as part of a CTD prevention program (see Chapter Nine).

Work Load Reduction

In the past, oversimplified approaches to improving productivity have sometimes resulted in some employees being simply overworked. Adverse effects include cumulative trauma, employee dissatisfaction, poor quality, and a net loss in productivity. Many companies have found that by reducing work load, cumulative disorders can be reduced, and sometimes overall efficiency actually increased.

For example, some meatpacking companies have increased staffing on some of the harder tasks. They found that a few more people resulted in cost savings from less wasted product and more than made up for the increased labor costs.

Team Building

Consideration should be given to provide a sense of belonging and being valued, particularly in the small work group. Barriers between groups and individuals may need to be addressed in order to provide a good understanding of other components of the overall work process. This concept can provide social support to employees and thus help reduce workplace stress, in addition to benefiting quality and workplace efficiency.

Training

A final area of work organization that deserves specific mention is training. There are many aspects of training that are relevant in preventing CTDs, including (1) encouragement of early reporting of symptoms as part of a medical management program and (2) education in the principles of ergonomics, in order to help generate ideas for physical improvements in the workplace. These issues are described more fully in the final chapter as part of setting up a workplace ergonomics program.

Training in basic work methods is also part of the effort. Two comments can be made in reference to jobs that entail a high risk for CTDs. The first has to do with the work method itself.

It is not uncommon to observe, where a number of employees are all doing the same task, quite a variety of specific work methods used by the employees. At first glance, they may look like they are all doing the same thing, since the steps of the job have been broken down and typically people are taught to do the task in that order. However, after closer analysis, people may be working quite differently regarding the little things. Some people can be working quite smoothly, while others move in a wide range of awkward motions. Good training programs can help all employees learn to use smooth work techniques.

The second aspect of work method training has to do with introducing new ergonomically designed equipment into a workplace, such as good chairs or adjustable workbenches. It is not sufficient to merely provide the equipment. People need training in how to use the equipment. Otherwise, the outcome too often is that of having spent money for equipment to help people, but it is equipment that they do not use.

THE VIDEO CAMERA—A GREAT NEW TOOL

In the past, job training meant teaching employees the basic steps of doing a job, not the nuances. Frederick Taylor aside, most people have been left on their own to learn the subtle techniques of a job. Typically, after a period of trial and error, most people finally learn how to do the task in the easiest way. Others, however, may never discover these techniques.

Fortunately, innovative approaches are possible. Video cameras enable employees to teach each other these subtle techniques.

The steps are simply to videotape various employees doing the job, then hold a meeting to have them see themselves in action. In most cases, people are totally unaware of their postures and actions while they work. By viewing the videos they can discover what they are doing. Often, merely by giving this kind of feedback improvements can be made.

Furthermore, by enabling people to see each other, they can learn from each other. Usually, employees are too busy doing their own jobs to watch what everyone else is doing. Also, some tasks happen too quickly and cannot be observed well on the spot. Watching a videotape, however, permits workers to see each other. The video can be put in slow motion and repeated, so that moves can be broken down into components and understood. Thus, a technology is now available to help in this regard that was not available a few years ago.

STUDIES ON WORK ORGANIZATION AND CTDs

With a few exceptions, this book contains no academic-style literature reviews, since the intended audience is nontechnical and most of the details are well documented in more technically oriented books. However, the relationship between work organization issues and CTDs is a new area and reviews of studies are not readily available elsewhere. Thus, a survey of the literature is provided.

The conceptual framework for linking work organization and cumulative trauma is (a) a variety of workplace conditions can trigger the physiological stress response, (b) which in turn contributes to CTDs by exacerbating the effects of the purely physical risk factors.

There are several major components to the scientific investigation of this topic. One is characterizing the physiological stress response. Another is understanding the types of workplace conditions that can trigger this response. A third is learning about the adverse health effects that are related to stress. A considerable body of

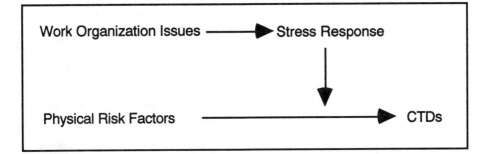

A simple conceptual framework for linking work organization and cumulative trauma disorders.

literature has been developed on these issues, and there is no need to summarize this literature here. Although there are many unresolved theoretical details, these aspects of stress have been well documented and are generally accepted.

The key issue is the link between the stress reaction and CTDs. The links between stress and many other types of disease have been established and the weight of the evidence is consistent with the suggestion that the stress response is a risk factor for CTDs. But the concept of stress and musculoskeletal disorders is relatively new and more investigation and review of existing studies on this topic are needed. One initial review has been conducted (Bongers et al., 1993) that begins to sort out the scientific merits of the studies that have been conducted thus far.

Studies on stress and CTDs fall into several categories: (1) those that link the stress reaction to CTDs, (2) those that link specific work organizational issues directly to CTDs, and (3) those that link stress with specific disorders, primarily back injuries. Finally, several consensus papers have been written on the subject.

It should be emphasized that, while the studies relating specific work organization factors directly to CTDs offer strong support for the general proposition, there is no scientific need to conduct studies on every one of these factors. The relationship between many of these work organization issues and triggering the stress response is known. All that is needed from a scientific point of view is evidence that the stress response contributes to cumulative trauma.

A Note on Causality

In the studies described below, associations were found between stress and cumulative trauma. From a strictly scientific viewpoint no causality can be assigned. Stress and CTDs are clearly related, but which causes which cannot be scientifically determined—as indeed whether both were caused by some other unknown factor.

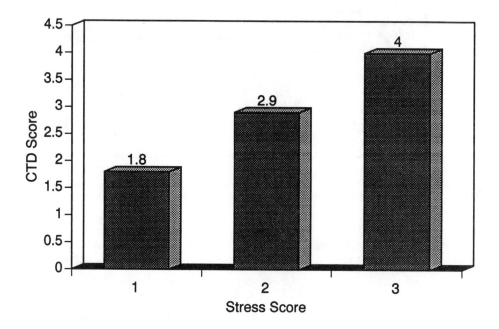

The relationship between stress and CTDs (adapted from Leino, 1989). The score on the vertical axis indicates the score for cumulative trauma—the higher the score, the more CTDs in the group of employees. The score on the horizontal axis indicates the degree of stress, based on a stress questionnaire administered to the employees—the higher the score, the greater the stress.

The researchers sometimes stated their own beliefs that the relationship is causal, but the data per se only showed an association.

Studies Linking Stress to CTDs

A variety of studies have directly assessed the relationship between stress and cumulative trauma. Perhaps the most ambitious effort has been made in Finland, using a group of 902 white-collar and blue-collar employees from a single facility monitored over a period of ten years. These employees were both administered a questionnaire assessing stressful conditions on the job, and given physical exams appraising signs of cumulative trauma. This study is unique in that it compared results over a period of time, rather than a single point in time. Solid relationships were noted: persons under high stress experienced up to twice the number of musculoskeletal problems than did persons with less stress. Furthermore, the study showed that baseline stress scores obtained in the first stage of the study served as predictors for eventual CTDs at the end of the study (Leino, 1989).

Another rather sophisticated study included complicating factors, such as the amount of physical risk factors to which employees were exposed and a variety of individual factors. After accounting for all of these variations, the basic relationship still held—high "psychosocial demands" are associated with problems in the back, neck, and shoulders. This study provided further support for the notion that muscle tension is an important pathway by which mental stress can affect musculoskeletal disorders (Theorell et al., 1991).

Several other studies have found similar associations in various industries. One occurred in an office environment, investigating the relationship between "psychosocial work environment factors" and physical ailments. An approximately threefold increased risk for neck and shoulder pain was found for those experiencing a "poor" as compared with those experiencing a "good" psychological work environment (Linton and Kamwendo, 1989).

A quite similar study addressed musicians in a symphony orchestra. Those reporting above-average stress experienced more musculoskeletal symptoms in the back and neck (Middlestadt and Fishbein, 1988).

One of the more intriguing reports addressed differences in CTD rates between teachers and interpreters in schools for the hearing-impaired. The interpreters appeared to have higher rates of wrist disorders than did the deaf teachers, despite the fact that the interpreters used sign language only up to four hours per day, while the teachers signed up to eight hours per day. The difference appeared to be related to the added mental stress involved in interpreting—the teachers were expressing their own thoughts, while the interpreters were simultaneously listening and translating (Cergol, 1991).

One especially well-done study for which I have a personal fondness investigated relationships between stress and several categories of insurance losses. Stress questionnaires were administered to several hundred employees in 40 different workplaces in the same industry and the results compared with insurance losses from those same workplaces. Higher levels of workplace stress were significantly related to higher workers' compensation costs (presumed to be related to CTDs), but not to vehicle accidents or other insurance losses. This evidence supports the concept of a *physiological* reaction, rather than a purely *psychological* one. If stress were to cause people to be distracted and accident-prone, the one would expect a relationship with vehicle accidents as well as workplace injuries. However, this was not the case—the vehicle accidents were not related, but workers' compensation costs were. Thus, workplace stress may create a response such as chronic muscle tension that predisposes the lower back to an injury when there is, for example, heavy lifting (MacLeod, 1988).

Studies Linking Individual Work Organization Factors to CTDs

The above studies were related to general measures of stress. Additional studies have addressed particular work organization factors and CTDs. These latter studies provide persuasive evidence concerning this whole topic, but as stated above there is no scientific need to prove a link for each specific issue of work organization and CTDs. Once a case has been made that a particular factor of work organization can trigger the stress response—and much of this is already provided for in the occupational stress literature—then it can be assumed that it can contribute to cumulative trauma.

Thus, the following studies provide further support for the general argument, but it does not follow that special studies need to be made for each and every work organization factor that is known to cause stress, but is not listed below.

One study in the telecommunications industry showed that employees in closely monitored jobs had higher rates of complaints associated with CTDs than did employees in less closely monitored jobs, even though the physical tasks were the same (Smith et al., 1990).

Four studies have linked issues related to job dissatisfaction to cumulative trauma. The first of these concluded as follows: "Subjects who stated that they 'hardly ever' enjoyed their job tasks were 2.5 times more likely to report a back injury than subjects who 'almost always' enjoyed their job tasks" (Bigos et al., 1991). The second study noted strong correlations between employees who reported work-related pain (back, neck, arms, and wrists) and having (a) feelings that the company did not really care about them, (b) less motivation to work, and (c) little say in their work (Dimberg, 1991). The third study concluded that many complaints of back pain could not be attributed to high peak loads, repetitiveness of lifts, or large loads. Monotony, stress, and low job satisfaction were considered to be more likely factors of greater importance (Magnusson et al., 1990). The fourth study showed that postal workers who had symptoms of upper limb CTDs also expressed greater dissatisfaction with machine-paced jobs as well as reported higher levels of tension (Arndt, 1983).

One study investigated the relationship between cumulative trauma and overall working relationships with others in the workplace. There was an eightfold higher risk of developing pain in the upper extremities if the working relationship was poor. The feeling of fellowship apparently played a more important role than the purely physical environment with regard to the risk of pain (Ryan et al., 1987).

Researchers at the National Institute for Occupational Safety and Health (NIOSH) have conducted a number of evaluations of workplaces to identify possible types of psychosocial and work organization variables that may be related to cumulative trauma. A number of specific factors were found to be related to CTDs, such as "increasing work pressure," "surges in workload," and similar items (NIOSH, 1990, 1992a,b). Researchers conclude, ". . . psychosocial and work organization factors need to be considered in programs to prevent cumulative trauma disorders; programs restricted to the physical environment are not likely to be fully effective" (Bernard et al., 1992).

Note again that causality cannot be established in these studies. For example, concerning the associations between job dissatisfaction and back disorders, the presumption of the argument is that the dissatisfaction contributed to the symptoms of back problems. But there are other possible interpretations. Perhaps it was the back problems that created the dissatisfaction with the work. Or perhaps the jobs were physically demanding, "back breaking"-type work that caused both back problems and job dissatisfaction.

Consensus Review

At least two consensus reviews have been prepared on this topic. The first relates to work with computer terminals with a conclusion that is quite relevant: "In the prevention of VDT-related health problems, psycho-social factors are at least as important as the physical ergonomics of workstations and the working environment" (Report on a World Health Organization Meeting, 1989).

Furthermore, the ANSI committee established to determine consensus and develop standards on cumulative trauma has accepted work organization as a risk factor, referring to stressful work situations (MacLeod, 1992).

Stress and Lower Back Pain

Quite a number of studies have been conducted on the subject of psychological factors in chronic low back pain. Typically, these are studies of individuals rather than working conditions, based on patients who report to clinics rather than employees in the workplace. It appears that patients with low back pain experience higher levels of stress, anxiety, and other psychosocial anomalies than patients without back pain. It is not necessary to cite and summarize those studies here, but one of these clinical studies did provide some unusual information and is worth reporting.

In this study, patients were given electromyography exams to measure levels of muscular tension in the lower back. The study found higher levels of muscle tension among patients with back pain than among controls. This muscular tension furthermore appeared worse under conditions of stress (Flor et al., 1985). Thus, there seems to be further evidence that the mechanism of action that links stress to CTDs is increased muscular tension.

Conclusion

These studies provide a considerable body of evidence linking (a) stress with CTDs and (b) specific work organization problems with CTDs. Most of the studies are quite recent, indicating that as interest in the topic grows, yet more studies will be conducted to establish relationships more precisely.

Two implications are:

1. Studies on the risk factors for CTDs in general should include a measurement of stress. By including this additional factor, many inconsistencies in the literature on the relationship between physical risk factors and CTDs may be resolved.

2. There is sufficient evidence to indicate to workplace managers that sources of workplace stress ought to be identified and eliminated or reduced where possible. The rationale for this action includes both CTDs plus other stress-related health problems.

STRESS IN NONINDUSTRIAL SOCIETIES

This section provides a different perspective on how to think about designing work systems. Anthropology can provide modern managers with many useful insights.

It is important to recognize that stress has not suddenly appeared with the advent of industry and modern society. People in hunter-gatherer and peasant societies undoubtedly experienced high levels of stress from sources such as the elements, human enemies, and the threat of hunting and crop failures.

Anthropologists have long noted the significant differences between primitive and industrial societies in certain patterns of work (Udy, 1959). We now recognize that these patterns can affect chronic stress. In particular, in nonindustrial cultures work is more typically characterized by short peaks of high activity followed by long period of low activity and rest. The short peaks may involve considerable stress, but the periods of low activity provide time for recuperation.

In hunter-gatherer cultures, this pattern in typified by the hunt. Stalking the game and the chase are "stressful," in that the body is geared for action and the physiological stress response provides the vigilance and adrenaline necessary to capture the prey. However, after the kill, there is a long, relatively leisurely trek back to the village in which the effects of stress are dissipated. Furthermore, hunter-gatherers physically work only a few hours per day, spending the bulk of their time sleeping, resting, talking, and engaging in purely social activities. Thus, despite the occurrence of stress during part of the working day, there is enough time available to recuperate from that stress.

Agricultural societies can also be characterized by patterned cycles of work and rest. Typically, high levels of energy are expended during spring planting and fall harvesting, but broken by long periods of less intense activity during summer and winter. Again, despite periods of high stress involved in completing work tasks,

OUR ANCESTORS RESTED

"The notion that Paleolithic populations worked round the clock in order to feed themselves now appears ludicrous. . . . As hunters our Paleolithic ancestors alternated bursts of intense activity with long periods of rest and relaxation."
Harris (1978)

there are other periods of less activity that allow recuperation and a return to a relaxed physiological state.

In contrast, work in industrial society can often involve almost constant levels of high involvement with stressors at work with little respite. To be sure, in industrial society some occupations have such cyclical swings, such as seasonal construction or farming. And nonindustrial work does not always follow the same pattern. Nonetheless, the general statement holds true.

If the genes of our ancestors were modified by thousands of years of these cyclical activities, those of us in advanced industrial societies may have inherited a biological requirement along these lines. Perhaps we should be designing our current organizations with these cyclical patterns in mind. Modern worklife provides the hunt and chase, but not the long walk back to the village. We have a biological inheritance that we cannot easily ignore. Perhaps the high levels of stress and the existence of cumulative trauma are the price we are paying.

WORKLIFE IN A MAYAN VILLAGE

In 1972, I worked in the Yucatan peninsula for the summer assisting in an anthropological study of Mayan worklife. The jungle village in which we lived was particularly unusual in that half of the villagers still supported themselves as hunters, while the other villagers depended solely on agriculture. Comparisons and contrasts were thus possible between these two types of economies, as well as with outside industrial society.

Work was hard, weeding crops under a searing sun and cutting back jungle undergrowth with a machete. The work was not always ergonomic, wielding hand tools repetitively from a hunched-over position. But the work only lasted a few hours, then people returned to the village, ate, and napped. People spent the rest of the day in conversation or puttering around their huts.

Long walks were part of daily worklife, either to the fields outside the village or to the deep jungle for hunting expeditions. (And walking is probably the only continuously repetitive activity that is good for people, no doubt related to the fact that for tens of thousands of years this is what people did, and a condition by which humans evolved.)

The summer rains were late that year, and there was considerable stress in the village as the cisterns ran dry. Yet the villagers shared the burdens equally. The people were closely knit and often related, and despite routine bickering and occasional long-standing feuds, the social support seemed to buffer the stress. I also observed a rain ceremony at a makeshift altar out in a field. The farmers were in this sense trying to gain control over the weather. It may have worked too well—the rains began a few days later and flooded the fields.

The ancient Maya had a better life. The nut from the breadnut tree, which grows naturally in the Yucatan, provided the substance of life in ancient times. The tree itself has deep roots that make it unaffected by the vagaries of rain. This nut provides high-quality protein and is effortless to harvest. The ease of production ("working smarter, not harder") gave the ancient Maya considerable time to invent culture and shape a sophisticated society that was in many ways more advanced than Europe at the time. Why contemporary Maya do not eat this nut anymore is a mystery and was the underlying reason for our study.

The experience provided me with two valuable insights on work organization:

Traditional Work Permitted Rest

The cyclical levels of work activities left a strong impression—peak work loads followed by periods of rest and recuperation. If this village represented the preindustrial life to which thousands of generations of our ancestors were accustomed, it suggests a biological criterion in how we ought to design work.

Human Organization Is Malleable

The most striking perspective that anthropology provides to modern management is appreciating how variable systems of human organization can be. We do not need to accept any culture-based aspect of work organization as "natural." We can feel free to set up any system we can think of to better meet our physical and mental needs. As we move into the twenty-first century we should not feel constrained because of cultural traditions in determining how to best organize ourselves. We can define human capabilities and limitations, then design systems to meet these requirements.

section 3

Case Studies and Workplace Programs

chapter EIGHT

SUCCESS STORIES

This chapter provides a series of case studies of successful
application of ergonomic principles. Several examples from
the author's experience in the workplace are included.
Additional case studies contributed by professional
colleagues show applications in product design.

CASE STUDY 1

Application: Manufacturing
Bottom Line: Workers' Compensation
Costs Cut Dramatically

This example shows that a one-time investment of $20,000 in a modest ergonomics program brought a return of $100,000 savings in annual workers' compensation premiums. The effort demonstrates a high level of creativity in using simple equipment in unconventional ways. This example is also useful because it illustrates a number of various principles and lessons of workplace design. Although this story happened years ago, it is still one of my favorites.

Background—Workers' Compensation Insurance Canceled

The facility was small, employing about 60 people, in a small town in a rural area. The company manufactured metal parts used in mainframe computers, taking advantage of the company's capabilities for high-tolerance machining. The plant consisted of two departments: (1) the machine shop, filled with millions of dollars of high-end machine tools, and (2) the deburring area, crammed with people standing at workbenches.

The company was experiencing high rates of cumulative trauma, both of the lower back and of the arms and wrists. The primary source of the problem was the deburring department; however, the machining department was affected as well. The costs became prohibitive and ultimately the workers' compensation insurance carrier canceled the plant's coverage. The plant finally was forced to seek insurance with the state workers' compensation pool. As the plant manager stated, "It's a little like having a D.W.I. Yes, you can get insurance. But, boy, do you have to pay for it."

Ergonomics Program

The company retained me as part of an effort to prevent the injuries and reduce the insurance premiums. I spent an initial two days at the facility, then returned one day per month for the following four months.

One of the first steps taken was to hold a meeting of the deburring employees to introduce me to the group, explain the program and ask for their cooperation in finding improvements. I gave a short presentation on the principles of ergonomics and CTDs, then the employees were given a chance to talk about problems they were having. It became clear as a result of this meeting that a larger percentage of the employees were experiencing problems—about 50 percent—than had ever reported any difficulties to management. Furthermore, the employees indicated the task that they considered the most difficult— deburring the "slotted pole."

Improving the "Slotted Pole" Task

Subsequently, one of the engineers and I reviewed the slotted pole task and discussed specific issues with the employee. The part was a cylinder about five inches long mounted on a stubby base. The inside of the cylinder was somewhat hollow, with three slots trisecting the part most of its length, hence its name. The job was to take the parts that had been recently machined and remove burrs and sharp edges. Because the part was to go into a mainframe computer, all slivers of metal had to be removed completely and the rough surfaces polished. Quality requirements were extremely high, since the slightest fleck of metal could ruin an entire computer.

The task of deburring the slotted pole was unpopular. Consequently a variety of people had been trained to do the work and rotated through the job to spread the burden equally.

One task consisted of holding the part in the left hand and manipulating it, while the right hand performed various motions such as sanding the part's edges

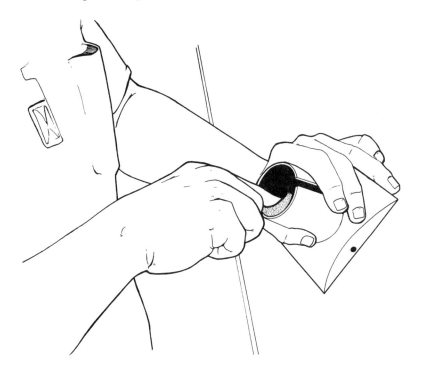

| Holding and sanding the part.

and surfaces. As shown in the previous illustration, the sandpaper was held in an awkward "pinch" grip with fingers extended to reach the inside of the part, thus creating high compression forces on the wrist of the right hand. However, the employee stated that the worst part of the task involved the left hand—the continuous gripping ("static load") and constant manipulation of the part.

Lesson 1:   It is vital to talk to the employee in order to learn about the job and gain insights into problems and solutions. Some details can be gained in no other way.

Lesson 2:   Using the nondominant hand as a "fixture" is a common source of cumulative trauma—there is a tendency to focus on the hand holding the tool and doing the work, and neglect the hand manipulating the product.

Lesson 3:   Often the specific problems of a job can be identified relatively easily, and without need for measurements or elaborate studies. Typically the difficulty is in finding solutions.

❙   Potter's wheel.

Based on this assessment and other observations, the engineer and I brainstormed options for improvement, and from time to time floated ideas past the employee and supervisor of the area. The first goal was to fixture the part to relieve the left hand from the constant gripping forces. The fixture would have to rotate, however, to accommodate the need to access the part in different orientations.

Initially, a simple lazy Susan fixture was considered. However, a concern developed that the need to continuously rotate the fixture manually with the left hand would merely transfer the cumulative trauma risk factors from the hand to the elbow and shoulder. Subsequently, the concept of powering the fixture was explored. This idea too eventually was discarded, since the slotted pole needed both to be rotated continuously and at varying speeds to sand the outside surface and then indexed rather precisely at one-third turn intervals to access the slots correctly. A power mechanism to accomplish both requirements was thought to entail excessive costs and development time.

Brainstorming then refocused on the lazy Susan, with the added idea of running a shaft from the rotating fixture on the workbench down the floor where it could be easily manipulated by the feet. As the engineer and I began to sketch the needed device, it occurred to us that a similar device had been invented long ago and was readily available.

The company purchased and installed a potter's wheel. Technology that was 4000 years old was used to solve a problem in a plant that made computer parts. There were many features of the potter's wheel that could be applied to this task. The potter's wheel had a plate, normally used to support the potter's clay, onto which a fixture to hold the slotted pole was attached. The shaft that connected the plate (and now fixture) to the wheel beneath could be manipulated easily by the feet, both to rotate evenly at varying speeds and to index precisely.

This off-the-shelf device worked successfully and served as the basis for further refinements. Specifically, the fixture eliminated all of the repetitions and exertion of one hand in manipulating the part (thus, in one step, reducing the overall wrist stress by at least 50 percent).

Lesson 4: Anything that solves a problem is ergonomics—the solution does not need to be a device normally thought of as an "ergonomic" product.

Lesson 5: The problem-solving process need not be elaborate or highly technical. Creative thinking is often more important.

Lesson 6: Unconventional, "harebrained" ideas often lead to good results. Solutions can be found in many areas, even from equipment not normally associated with the industry in question.

Lesson 7: Fixtures, where feasible, can be quite effective in reducing stress on at least one hand.

Another part of the task involved use of a generic "deburrer's knife." As shown in the illustration on the next page, this knife was used to reach deep into the center

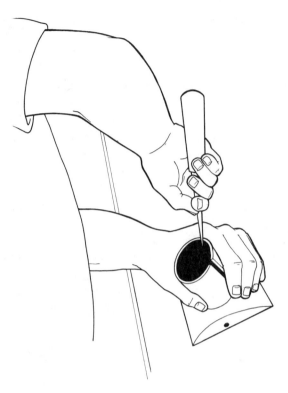

| Deburrer's knife.

of the cylinder to scrape out burrs. The illustration shows several other issues. One is the awkward postures of the wrist and elbow, both contributors to cumulative trauma. Another issue is the grip of this generic device, which was so inadequate to the task that it was not used at all. Finally, the illustration shows the problem with the left hand, an extended pinch grip that was eliminated with the use of the fixture as described above.

Once the part was attached to a fixture, the left hand became available to help manipulate tools. In this case, a special two-handed tool was designed specifically to reach into the slots to scrape out burrs. The tool could simply be inserted into the slot and pulled, providing three great benefits: (a) the tool placed the wrists and elbows in much better posture, (b) it divided the force between both hands, and (c) it took advantage of the larger muscle groups in the upper arms and even shoulders and torso.

Lesson 8:   Once a product or part is fixtured, it opens the door to two-handed tools, which usually are both easier to manipulate and distribute force to more (and often larger) muscle groups.

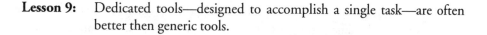

I   Custom, two-handed knife.

Lesson 9:   Dedicated tools—designed to accomplish a single task—are often better then generic tools.

A further problem in the original task was sanding the outside of the part. Considerable repetitive, rather forceful motions were required to complete the task. Once again, the left hand was used as a fixture in doing this step of the job.

The potter's wheel purchased represented the high end of potter's wheel design, and came equipped with an electric motor to spin the wheel. As applied to this deburring task, the wheel could be manipulated by the feet in order to index and move the part slowly, but the motor could also be turned on to spin the wheel. In this mode, the employee could simply hold a piece of sandpaper rather loosely to the part and it would be polished quickly and cleanly with little physical effort.

Lesson 10:   Reducing repetitions does not necessarily mean slowing the job down. On the contrary, with good design, the task can be completed faster, but with less manual repetition.

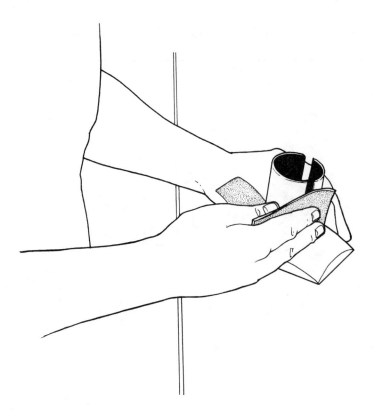

| Hand sanding, before.

As experience was gained with the potter's wheel, a variety of modifications were made. The plate (to support the potter's clay) was removed, since it had no function anymore. The fixture was angled with the use of a gearbox to orient the slotted pole to the employee for improved access. A storage area that was part of the original equipment, but a bit out of reach, was raised up and moved closer. A task light was added. The original potter's stool, which had no back support and lacked cushioning for the seat, was replaced with a car seat, equipped with an electric motor to provide adjustment for varying leg length.

Lesson 11:   Continuous improvement applies to ergonomics as much as to any other aspect of the workplace. There is no final ergonomics fix— there are always ways of improving equipment.

Two additional items of special note also evolved. The first involved an additional problem with the initial potter's wheel design. Once the wheel was spinning, there was no way to stop it quickly. Potters had no particular need to stop the wheel as abruptly and frequently as did these deburring employees.

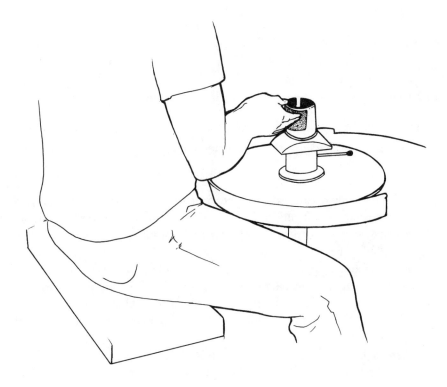

| Hand sanding, after.

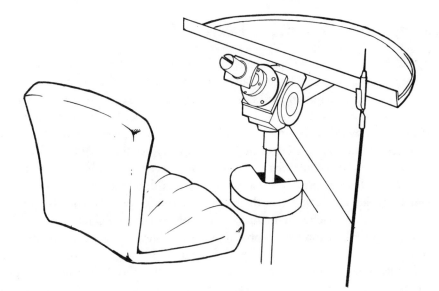

| Modified potter's wheel.

The solution involved the plant maintenance man, who in the course of events had become actively engaged in redesigning the potter's wheel. He happened to be a Harley-Davidson enthusiast and had a supply of spare motorcycle parts available. To stop the wheel, he attached a Harley-Davidson brake, complete with foot pedal that stood out conspicuously to one side—large and chrome-plated.

Lesson 12:   Everyone can make a contribution. Maintenance personnel, once provided with the principles of ergonomics and the goals of the program, can be especially gifted in making creative improvements.

On previous occasions, the supervisor had encouraged the employees to use various powered deburring tools that were available. However, the employees resisted, stating that the powered tools were awkward to use. As it turned out, once the employees began to use the potter's wheel, they also began to use the power tools successfully. Apparently, the employees had become so used to doing the task in a certain way that any change, even ones that were seemingly beneficial, seemed awkward. However, once the entire task was disrupted and whole new techniques needed to be learned, the employees were able to incorporate the power tools.

Lesson 13:   Incorporating change can be difficult for a variety of reasons, one of which is the chore of learning a highly refined work technique all over again. Sometimes it is easier to make a big change than a little one.

Improving Other Deburring Tasks

After the initial changes were made on the slotted pole, attention shifted to other tasks, again involving employees and brainstorming improvements.

Most other deburring tasks involved working on large, traditional worktables. Employees often worked hunched over these tables, with poor back and neck postures. The workbenches were fixed-height and could not be adjusted for taller workers. Moreover, since two or more people typically worked at each table, the tables could not accommodate everyone easily, even if the tables had been adjustable. An additional issue was that the employees stood all day on a concrete floor. Finally, other issues were that there was no particular place provided for tools, plus occasional long reaches to obtain needed tools and equipment.

The first decision was to procure sturdy, adjustable-height worktables. However, none could be found on the market when this case study took place in 1987. Happily, once again with creative thinking, an off-the-shelf product—originally intended for another purpose—was modified for use, rather than constructing a suitable table from scratch.

The item purchased was a standard industrial die cart—designed originally to

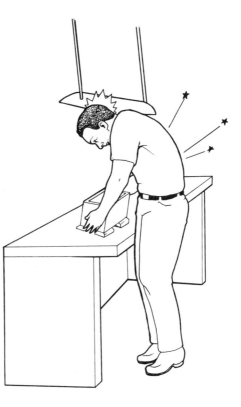

| "Before" workbench.

lift heavy dies into presses—that had many of the features needed, particularly height adjustment and a sturdy base. It was modified in several ways. A quarter-inch sheet of plastic was attached to the top surface to provide more of an appearance of a worktable and to protect tools and parts from the steel. Brackets were mounted to the sides of the table to hold tools, fixtures, and a task light.

Used as a workstation, these carts could readily be raised and lowered to adjust for individual height. Furthermore, the die cart had several additional features that unintentionally came into play:

- The base of the cart served well as a footrest, whether sitting or standing.
- The area of the top work surface was small, only about three feet wide and two feet deep—quite sufficient for the task at hand. The unintended benefit was that once all employees were equipped with these tables and the old, larger tables discarded, considerable floor space was gained. A previously congested area became rather roomy.
- The die carts came with wheels, since they needed to be moved about in their original application. These new workstations were thus mobile and could be used as the basis of a modular, flexible manufacturing system. The engineers created several work cell areas using pedestals attached to the floor and

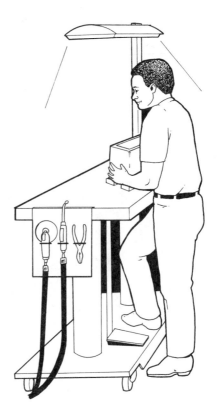

"After" workbench.

equipped with electric outlets and pressurized air couplings. Employees then wheeled their workstations to the appropriate cell, hooked up and worked as a unit on a particular part. As production required, they shifted quickly from one cell to another.

- These individual workstations could be personalized for each employee to promote an individual's identity.

Finally, the workstations were designed for standing height, and tall stools were provided to permit employees to sit or stand as they so chose. The stools themselves were adjustable and had other ergonomic features. For standing, antifatigue mats were provided.

Lesson 14: Once each employee is provided with an individual workbench, individual adjustment becomes feasible.

Lesson 15: There were unanticipated benefits. Many tasks do not require the space allotted to them. Workbenches and desks are often designed to be large for no reason other than status, and they merely serve to collect clutter.

Improvements in the Machine Shop

Although the machining area was not the focus of initial efforts, eventually engineers identified ways to make improvements here as well. For example, an articulated arm, previously purchased for another project but found to be infeasible for that task, was successfully used to lift the heavy parts in and out of the machine tools. Additionally, an ingenious conveyor line was constructed to move parts from one machine tool to another instead of loading onto a cart, moving a few feet, then unloading the cart. Thus, considerable time-wasting and stressful repetitive motions were eliminated.

Postscript

As a side note, engineers in facilities such as this often gain their personal satisfaction from machining the metal, employing expensive machine tools to create high-tolerance parts. Deburring is a necessary evil, an afterthought and not of real interest. The deburring area of this plant, as elsewhere, evolved without much engineering support.

However, in this facility the engineers now recognized deburring as a problem in a way in which they had not previously, and subsequent to the ergonomics project found ways to improve machining capabilities to reduce the burrs produced and thus eliminate much of the need for deburring.

Lesson 16: Often, just getting people to think about a problem, perhaps presented in a new way, can be sufficient to induce them to apply their usual skills in solving problems. Extensive training in the details of ergonomics may not be necessary.

Lesson 17: One should not neglect the root cause of the problem.

Results

Several positive results emerged from these changes:

- Workers' compensation costs dropped 60 percent during the following year. Cumulative trauma cases were prevented and within a year the company was out of the state insurance pool and able to obtain normal insurance.
- Employee satisfaction increased. The employees clearly appreciated the new workstation. In particular, employees complained whenever the potter's wheel was taken out of service to be modified, forcing them to do the job in the old way. An additional potter's wheel ultimately had to be purchased as a standby.
- Productivity increased. Using the new workstation, employees were able to produce six parts per hour compared to the previous five per hour. (However,

the work standard was not changed since the management's focus was on quality rather than quantity produced. Furthermore, the managers thought that continuing to produce only five per hour would permit additional time for rest, which would further reduce the risk of cumulative trauma.)

- Quality improved. With the workstation changes and the extra time to do the job, costly batch rejection rates were reduced to nearly zero.

Costs and Benefits

The costs of improving the deburring area amounted to a one-time investment of about $20,000, including both administrative time and purchase of equipment. The savings in workers' compensation premiums during the following year alone were $100,000.

CASE STUDY 2*

Application: Military Weapons System
Bottom Line: Faster Deployment;
Better Operator Communications

This case study describes a systems approach to designing a military ground station. The procedures used to develop the physical layout for this unit apply when designing all types of human–machine interactions.

The Design Challenge

The U.S. Army needed a small highly mobile ground station to house a two-person crew. In this station they received and analyzed electronic data, communicating their interpretations to other military units. Unfortunately, the original layout did not adequately support the needs of the crew inside, especially under the stress of the modern battlefield.

The design challenge was to allow crew members to work as a team and share data. Obstacles to that goal were cramped, claustrophobic quarters, which increased stress and reduced effectiveness.

The Systems Process

Understand System Functions

The first step was to define the role that the ground station was expected to play within the Army's command and control structure. Ergonomists reviewed Army

*Contributed by Frank Foss, FC Foss Associates, Inc.

documents describing the ground station's mission and requirements to develop a *task analysis.*

The second step was to continuously evaluate and revise those findings. This proved critical to decisions made concerning design, training, and performance assessment.

The third step in this process was to develop a *system network diagram.* This diagram became the basis for later network simulations, operational analyses, task groupings, and job design.

Understand People

By understanding the skills, knowledge, and abilities of the ground station operators, ergonomists could then allocate functions, match people or machine by tasks, and capitalize on the behavioral strengths of each. For instance, humans are not very good at performing repetitive tasks over extended periods of time, such as picking targets on a moving display. Computer software, on the other hand, could easily track targets if someone first identified which "blip" to follow.

By combining the creative abilities of people with the tireless processing power of computers, it was possible to create an automatic tracking feature. This gave operators the opportunity to develop target information that would not otherwise have been possible.

Understand Conditions

Ergonomists conducted a number of field observation studies during military maneuvers. These findings were reviewed and augmented as needed to define the working conditions of the ground station crews. The psychological and social environment received special emphasis, along with the constraints imposed by military regulations.

Define the Design Problem

Computer simulations were used to assess the alternative information networks, the ground station configurations, crew structures, and tasks. Based on these findings, the designers prepared detailed specifications, describing the functions to be performed, the human or machine components performing them, and performance standards. Furthermore, these findings indicated that the activities of the crew and, hence, the design of the ground station would change drastically, depending on where the ground station was used within the Army's command network. As a result of this analysis, military planners revised and clarified the ground station's mission objectives.

Create Initial Concept

A focus group was created comprised of the ergonomics design team, hardware and software engineers, shelter designers, and user representatives. Led by an experienced focus group leader, the team brainstormed and formulated design alternatives, to develop a *component interaction matrix.* This matrix paired each component with every other component and then scored the frequency and importance of operator movements between components. Frequent movements were allotted a high score, with infrequent movements scoring low. The components' interactive scores were then ranked.

To ensure that the fundamental physical relationships between controls, displays, and other critical elements were placed within an operator's field of vision, a component "bubble drawing" was created. The drawing is a graphic technique in which bubbles symbolize the components and are arranged in order of importance outward from a center, with less frequently used elements placed around the periphery. Operator movements, illustrated by crossover lines among the bubbles, were minimized to ensure smoothness and fluidity of motion.

The component bubble drawing was the basis for more than a dozen *physical layout sketches,* each representing a slightly different mission concept. These sketches were used by military planners to understand the significance of alternative ground station deployments, and by designers to explore variations in component placement.

Develop Prototypes

Based on an optimal physical layout sketch, engineers constructed a full-scale mock-up of the ground station from inexpensive materials, with critical ground station characteristics. This prototype could then be easily reconfigured visually and functionally to evaluate evolving ground station designs.

Test Prototypes and Revised Designs

Both civilian and military test subjects empirically tested the ground station designs using the mock-up. Special attention was given to the subjects' ability to perform typical mission tasks, as well as access components for installation, servicing, and maintenance. A number of adjustments were made in the layout to better accommodate

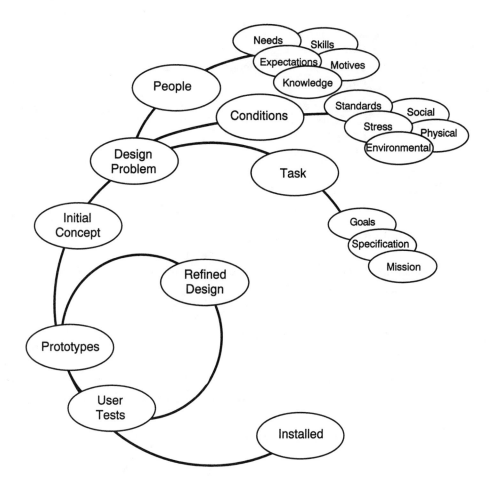

"Bubble drawing" showing components of the system arranged in order of importance, beginning at the center.

available hardware and to improve accessibility for maintenance. In addition, prototype testing suggested less immediate improvements that, although not incorporated into the current design, were recorded for use in future developments.

New Layout

Several key features of the final ground station design were:

- adjacent display consoles angled at 16 degrees, allowing operators to easily view each other's displays

- inclusion of adjustable analysis work surfaces consisting of two large digitizers on which maps could be mounted
- easy access to both of the operator positions
- clean smooth surfaces
- carefully placed lights
- conveniently located environmental controls to regulate temperature and ventilation

Outcome

1. Improved feeling of spaciousness.
2. Improved feelings of personal control and well-being.
3. Improved operator's ability to compare and discuss imagery received from different sources.

Final Ergonomics Payoff

Many of the concepts generated through this project were implemented into a similar version of the ground station used effectively during the 1991 Persian Gulf War.

CASE STUDY 3*

Application: Design of Photocopying Machine
Bottom Line: Drastic Cut in Service Costs

Reducing the cost of doing business justifies ergonomics in product development, but it often comes too late to prevent expensive mistakes. The need for proactive ergonomics input into product design soon becomes apparent and easily cost-justified if records are kept of everything associated with poor workplace design. This includes accidents, near-accidents (incidents), equipment service demands, along with records of customer complaints and returns. An incident used to cost-justify an ergonomics specialty in the design and development of copy machines illustrates this.

Every time a company service technician responded to a call for copy machine service, the problem and repair solution were described on a prepared form and sent

*Contributed by Richard L. Patten, 3M Company.

to company headquarters for analysis. About 35 percent of the service calls were for "operator-correctable" (O-C) problems, caused by the user's failure to operate or attend to the machine in a correct manner. For example, if a copier user did not remove a piece of jammed copy paper from the paper path of the machine, the service call would be recorded as an O-C. If they did remove the paper jam, but were then unable to restart the machine, it *might be* an O-C, depending on the cause of the problem.

The justification analysis continued, examining the service call records for each machine model. We selected and analyzed those with a high percentage of O-C service calls. We reviewed the service record descriptions of the problem, as well as physically examined the machines.

One copier had a high frequency of service calls for a problem with the toner-developer assembly. Customer described the problems in various ways, such as:

- "won't copy"
- "toner mechanism doesn't work"
- "I removed a paper jam behind the toner unit, put the toner unit back in, and then it wouldn't work any more."

The service records indicated that the power plug for the toner-developer unit (which distributes the black copying powder/toner onto copy paper) was found disconnected. Simply reconnecting the power plug fixed the problem. The service call was classified as O-C because the user did not follow the correct procedure of reconnecting the power to the toner unit once the paper jam was removed. An ergonomics review of the machine design *prior* to production would have caught this elementary design error.

Fortunately the company division involved in the cost justification described here was also responsible for both product development and service. Not all companies or company divisions have an accounting system sophisticated enough to support cost justification for changes in product design or the design method itself. And in some companies, service is associated with sales units whose budget and management units are separated from the product development unit.

In this case study, the costs of design and service were items in the same balance sheet. Division management was able to see that a preproduction ergonomics review of the machine design would have caught this elementary design error. Accounting's solution had been to simply increase the charge for service calls. But by spending more on the initial design process and adding an ergonomist to the design review team, the company saw an overall increase in their profitability plus improved customer satisfaction.

Outcome

- The estimated savings from the cost of servicing this one O-C problem over two years of the study were about $500,000 a year.

- Subsequent machines were designed so that the toner unit did not block access to the paper path. When the unit was moved (to add toner, for example), electrical connections were unaffected, or were automatic rather than manual.

Final Considerations

The legal and medical units of a company may have budgetary management separation from product development and production units. Ergonomic improvements that reduce medical costs may not be justified as reducing production costs. Likewise, since litigation costs are not usually part of the product development budget, it is sometimes difficult for development managers to see how they can improve their unit's bottom line by spending more at the outset for safety-related work by human factors specialists.

Thus, the first step toward cost-justifying an ergonomics specialty in your company may well be one of uncovering the costs of poor product or workplace design and bringing those into the same budget as product development. If we learn anything from such an approach, it is that the often hidden costs for product service, returns, and recalls (not to mention the more devastating litigation and medical and workers' compensation costs) can all be averted.

CASE STUDY 4

Application: Meatpacking—Ergonomically Improved Boning Line
Bottom Line: Increased Yields and Reduced Workers' Compensation Costs

A major design change was implemented in one facility for the method and equipment used for boning meat products. Many of the features of the equipment are quite unconventional and are proprietary to the company.

Introduction

In this facility, analyses of OSHA 200 records and workers' compensation costs indicated that the boning line was a priority. Furthermore, initial task analyses and brainstorming meetings indicated that improvements were feasible.

In the United States, most meat products are boned assembly line style—actually on a "disassembly" line. The animal is first killed, gutted, split, cleaned in a variety of ways, then chilled. After chilling, the carcasses are sawed into quarters or primal cuts, which are subsequently sent down separate lines to be deboned, trimmed, sorted, and packaged.

On this line, each step was analyzed for CTD risk factors and other ergonomic issues. Options for improvement for each issue were brainstormed in a team setting. Among other planning steps taken, an experimental table was fabricated to enable employees to test various heights, tilt angles for cutting boards, and layout configurations.

The features incorporated into the improved line are as follows. Brief descriptions of traditional lines are provided to permit comparison.

Heights

Traditional Lines

A single-height conveyor line is used for all tasks. With approximately 15 employees along this boning line, the line is above waist height for shorter people, and at knuckle height for taller people.

Improved Line

Employees stand on separate, powered adjustable-height stands operated by push-button controls that permit employees to adjust for their own height, plus change throughout the workday.

Furthermore, through experimentation it was found that various tasks were best done at different heights, for example, removing fat at above the elbow height and removing the bone at below elbow height. Stands were thus designed to put the employee at the best approximate height to perform the task.

Standing/Sitting

Traditional Lines

Employees stand continuously for the entire shift, either on hard concrete (or tile) floors, or on grating that typically is rather rigid.

Improved Line

Each workstation is provided with a swing-out support that holds either a lean-stand or a sit-stand, interchangeable at the employee's discretion. Thus, employees are allowed the options of sitting, leaning, or standing.

Furthermore, grating is mounted on 3/4"-inch springs, which provides notice-able cushioning.

Repetitive Arm Motions and Reaches

Traditional Lines

As muscles, bones, and trim are removed, they are typically either: (1) tossed up onto conveyor belts mounted overhead (these conveyors can range from one or two belts located at shoulder and/or head height, to four or five conveyor belts located upwards to three feet overhead) or (2) tossed into bins located behind employees, involving an over-the-shoulder toss. Repetition rates vary, but can reach about 200 tosses per hour (once every 18 seconds).

Improved Line

Conveyors are located to the side or beneath the work surface. All product is thus dropped downward from a maximum of an arm's length from the cutting area. All overhead or over-the-shoulder tosses are eliminated.

Traditional Lines

Depending on the design of the line, muscles are skinned of membrane and sorted by type, either by tossing into overhead conveyors as described above, or by tossing into a series of bins located in various positions around the skinners. In the latter system, the tosses range from 2 to 15 feet laterally, at a rate of about 450 tosses per hour (7.5 tosses per minute, or one toss every 8 seconds).

Improved Line

All completed products are dropped into a hole in the work surface located about 4 to 6 inches from the cutting area. The product falls onto belts that convey the product into bins. Product is sorted or combined by use of a series of gates on the conveyor belts.

Repetitive Wrist Motions

Traditional Lines

Much of the wrist stress occurs on the nondominant hand, which is used to hold and manipulate the product.

Improved Line

The product is supported by a simple fixture, a hook—sometimes called a "third hand." The hook virtually eliminates repetitive motions in the nondominant hand.

Hand Exertion

Traditional Lines

One hand is needed to hold the product, while the other hand is used to hold the cutting tool.

Improved Line

Fixturing the product to the cutting board allows alternating between left and right hands or using both hands to hold the cutting tool, thus reducing static load and exertion (as well as repetition).

Traditional Lines

The nondominant hand is used to hold the product with a pinching motion of the extended fingertips.

Improved Line

Fixturing the product enables use of a hand-held meathook to pull muscle away from bone. (The meathook itself needs improvement to prevent discomfort to the hand, but even traditional designs of the meathook improve on use of the gloved hand alone to grasp the product.)

Traditional Lines

The meat product is provided to the boning lines directly from chilled vats.

Improved Line

The product is run through a tumbler for 2 to 3 minutes, which softens the meat and enables easier boning.

Back, Neck, and Wrist Posture

Traditional Lines

Traditional cutting boards are horizontal, such as with a normal table or workbench. Consequently, on a boning line, many employees work with neck and back deviation greater than 25 degrees.

Improved Line

The cutting boards on this line are tilted, which permits neck, back, and wrist posture to be closer to neutral positions. Furthermore, the angle of the tilt changes depending on the task. Thus, the equipment automatically presents the product to the employee in an optimal position for the specific task.

Noise Control

Traditional Lines

Noise sources are above 90 dBA, requiring hearing protection.

Improved Line

A variety of noise-dampening features were included in this design. Furthermore, in this case, the line was located in a separate room and thus isolated from other noise sources. Thus, employees may work in this area without use of hearing protection. Layout of workstations also allows employees to converse readily with each other.

Work/Rest Schedules

Traditional Lines

Fifteen-minute breaks are provided every two hours.

Improved Line

A one-minute "microbreak" is provided every hour, simply by allowing the first employee on the line to stop placement of meat on the line for a minute. A break is thus achieved with minimal disruption. The break simply moves its way through the line, allowing employees to stand back, stretch, and relax, without leaving the area. These breaks occur in addition to the usual 15-minute break.

Planning Process

Traditional Lines

Plans are made by engineers and operations management with little or no discussion with employees or outside expert.

Improved Line

Ideas for improvement were developed in a team process involving suggestions from employees, operations managers, plant engineers, equipment vendors, and me (the outside expert) in an interactive, brainstorming process.

Results

To date, the evaluations show:

- CTD risk factors were reduced, as described above.
- Workers' compensation costs were reduced.
- Production yields were increased.

(Specific details are proprietary to the company.) Note that although the benefits of this line over the traditional lines are evident, this case study does not imply that the concept is the ultimate solution. Undoubtedly, better ways yet to perform these boning tasks will be found.

CASE STUDY 5

Application: Paper Processing
Bottom Line: (a) Finger Repetitions Reduced from 50,000 to Near Zero; (b) Productivity Doubled

A particular beauty of the following case is that an employee developed the ideas. Her formal exposure to ergonomics was only slight—a one-hour training session held at the plant, plus reading and hearing about the ergonomics program in the company.

Task   Receiving department; hand slip-sheeting pallets of paper
Task Description   The task involved counting stacks of folio-sized paper and inserting a slip sheet every 50th or every 100th sheet. To accomplish this, the employee manually counted the paper in groups of 25, lifted the group of paper from one pallet to another, then placed a slip sheet on the counted groups as appropriate. The pallets were placed on the floor and a pallet jack was used to transport the product.

Ergonomic Issues

Hand/wrist	45,000 to 50,000 finger repetitions per day in dominant hand to count through the paper.
	Static pinch grip in nondominant hand to hold group of paper while counting.
Arm/shoulder	Awkward, static postures of both arms to hold paper while counting.
Low back	Often working bent over to reach the lower half of the pallet.
Morale	"No one wanted to do this job."

Improvements

1. A semiautomatic paper counter was purchased. Rough groups of paper could be slid into the counting machine, then slid out in precisely counted groups.
2. A small, adjustable-height lift truck was purchased to support the incoming pallet of paper. This lift truck was used both to transport the pallet of paper and to raise the pallet to a comfortable height.
3. An "air table" was installed to hold the groups of paper to be counted. The paper could now be slid, rather than lifted. The air table consisted of a standard 30-inch-high table with pressurized air forced through small holes in the work surface, which thus minimized friction of the paper pulled across the table surface. The air table was recycled from another operation that had been remodeled and thus was cost-free.

4. The above equipment was placed next to an unused scissors-lift table that had previously been recessed into the floor. A special counting work area was thus created.

5. Control switches for the lift truck were mounted on an extension cord to reduce long reaches while counting the paper.

Results

1. Finger repetitions were reduced to near zero. (A small percentage of extra-thick paper still had to be counted by hand, since it did not fit into the counting machine.)

2. Lower back, arm, and wrist postures were all improved. The employee was now able to work in good "neutral" postures for most of the day.

3. Static postures were eliminated.

4. Repetitive lifting of counted groups of paper was eliminated.

5. Exertion to manipulate groups of paper was reduced.

6. Productivity doubled.

Cost

Paper counting machine: $16,000
New lift truck: $3,000

Savings

Labor: $15,000 per year
Quality: More accurate counts
CTDs: Drastically reduced risk factors

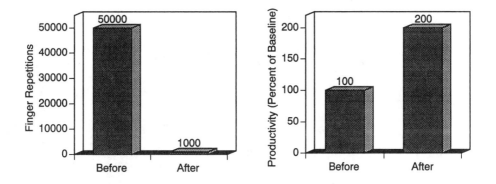

Finger repetitions were reduced from 50,000 to near zero, while productivity doubled.

CASE STUDY 6

Application: Metal Manufacturing
Bottom Line: (A) Injury Rate Reduced from 24 Percent to Zero;
(B) Productivity Increased 9 Percent;
(C) Absenteeism Dropped 1.4 Percent;
(D) Discomfort Reduced 31 Percent

Background

The following approach was taken in a manufacturing facility of 1000 employees:

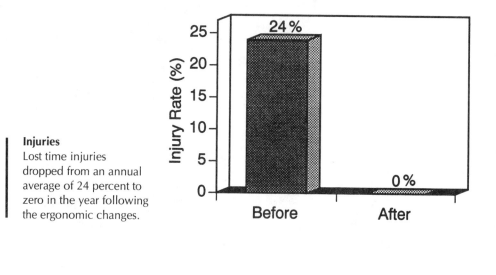

Injuries
Lost time injuries dropped from an annual average of 24 percent to zero in the year following the ergonomic changes.

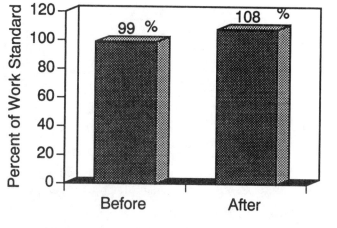

Productivity
Labor efficiencies improved 9 percent from an average of 99 percent of standard to 108 percent. This alone translated into a payback time of less than six months for the ergonomic improvements.

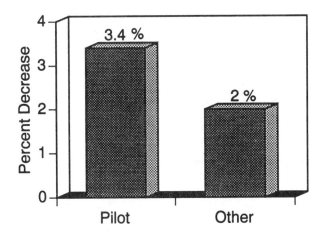

Absenteeism
Absenteeism in the pilot department dropped 1.4 percent over and above the plant improvement of 2 percent for the same time period.

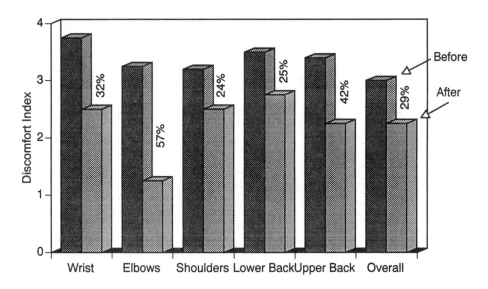

Discomfort
Physical discomfort was reduced an average of 31 percent. Employees were surveyed for symptoms of aches and pains using a discomfort index before the ergonomic changes were made, and then six months after.

Injury and Cost Analysis

Injury rates and workers' compensation costs were analyzed to identify a high-injury department for a pilot project. By a variety of criteria, one particular department surfaced as a critical area.

- *Training*—Instruction in basic principles of ergonomics was provided to a number of groups in the facility:
  - 10 hours for engineers and tool builders
  - 3 hours for supervisors
  - 2-hour briefing sessions for top plant management
  - 1-hour session for hourly employees in the pilot department
- *Organization*—We selected a cross-functional ergonomics team to evaluate in detail the pilot department and recommend specific solutions. Several hourly employees were included in the team along with managers, the department engineer, and the company safety director.
- *Workstation evaluations*—I reviewed each workstation and discussed the concerns with each employee. Because of the high variety of tasks performed, more than I could observe, heavy emphasis was given to the employee viewpoints. The brief training session in ergonomics provided enough information for the employees to make insightful suggestions.
- *Discomfort survey*—Employees were given an anonymous self-administered questionnaire that asked them to rate the amount of physical discomfort they felt in various parts of their bodies. Results were used to help pinpoint problem jobs and to serve as a baseline to evaluate the effectiveness of the improvements.
- *Improvements*—Changes included innovative design of workbenches, special tools, and work areas. Additional improvements included hydraulic scissors lift stands, parts stands, adjustable-height workbenches and chairs, improved layout, and improved task lighting.

CASE STUDY BONUS: FOUR QUICK FIXES

Application: Various
Bottom Line: Improvements
for Little or No Costs

Example A	Machine operator, working on bench-mounted machine.
Ergonomic Issues	Bent neck, bent wrists.
Improvement	The back legs of the machine were mounted on a small block of wood, thus tilting the machine forward. Both the wrist and neck postures were simultaneously placed in improved posture.
Cost	None.

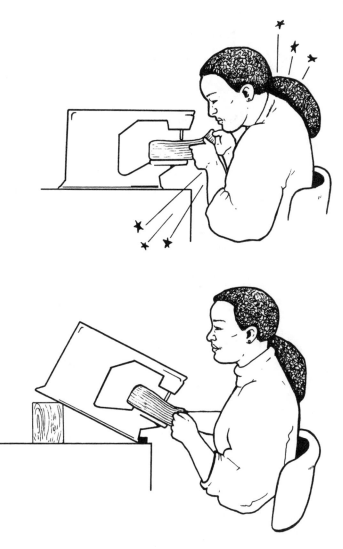

Before and after of necks: a simple block of wood solved the problems.

Example B

Ergonomic Issues

Improvement

Cost

Meat cutting with knife.

Highly forceful arm motions with outstretched arm and bent wrist.

For certain cuts, the front end of the cutting board was mounted on removable stops, to permit the board to be tilted 45 degrees down and away from the butcher. Thus, the butcher could press down, rather than forward, taking advantage of gravity and larger muscle groups. Additionally, the arm and wrist were placed in better posture.

< $100.

Example C	Meatpacking—removing organ meats with knife.
Ergonomic Issues	Constant static grasping force to hold knife. Repetitive hand motions for both knife hand (cutting) and nonknife hand (grasping product).
Improvement	A fixture was built and mounted to the work surface to hold the knife. This fixture was innovatively designed for quick exchange of dull knives for sharpened ones (five to six times per shift), plus had appropriate safety guarding. The product could then be grasped with both hands and easily pulled through the fixtured knife.
Cost	< $50.
Example D	Manufacturing plant—assembly of final product.
Ergonomic Issues	Repetitive lifting of product from main conveyor line to workstation, then back onto conveyor.
Improvement	A six-inch length of roller conveyor was added between the workstation and the main conveyor, allowing the product to be slid rather than lifted.
Cost	< $50.

N INE chapter

A GUIDE TO WORKPLACE PROGRAMS

This chapter provides an overview of administrative issues that typically need to be addressed when setting up a workplace ergonomics program to prevent CTDs. In fact, these issues comprise the elements of the proposed federal OSHA regulations and other guidelines and standards in this area. In many ways, this chapter represents how I think a standard ought to look, along with comments on each section.

The chapter outlines the basics of a workplace ergonomics program, plus contains tips and perspectives to empower this process. However, it is not a detailed "how-to" document, which is available elsewhere (MacLeod et al., 1992, 1993).

The focus of the programs described in this chapter is the prevention of CTDs. However, the same process can be used for other ergonomic issues as well, such as improving product quality or workplace efficiency.

PROGRAM ELEMENTS

Every organization concerned about high workers' compensation costs and cumulative trauma, regardless of size or industry, should adopt an ergonomics plan as part of its normal business operations that addresses the following basic program elements:

1. *Organization*—a plan for who should do what and by when.
2. *Training*—an effort to provide training in ergonomics to people at all levels of the organization.
3. *Communication*—mechanisms for the personnel involved in the process to exchange information about activities and progress.
4. *Job analysis*—a systematic process to review all jobs for issues and to identify possible improvements.
5. *Making job improvements*—the key part of the process, changing tools and tasks to be compatible with human limitations.
6. *Medical management*—procedures and protocols for identifying and treating employees with symptoms of cumulative trauma.
7. *Monitoring progress*—ways to measure and evaluate the program.

These program elements are described in more detail in the following pages, along with tips for setting up and empowering your program.

Comments:
| **Program Elements**

If a company wishes to cut workers' compensation costs, these are the basic issues that must be addressed. The above elements are the "standard," that is, what is expected of every employer. Everything in addition to this standard should be considered an appendix.

These elements are broad enough to be applicable for every size company in any business, from a huge meatpacking plant to a small mom-and-pop grocery store. In small organizations, these program elements can be achieved with relatively modest efforts. In large organizations, or ones with severe problems, considerable work may be necessary to implement a good program. Despite these differences, this framework provides a useful approach for thinking about ways to set up programs.

These program elements constitute the basic components of existing guidelines and proposed standards in this area (that is, OSHA's *Ergonomics Program Management Guidelines for Meatpacking Plants,* OSHA's expected regulations *Ergonomic Safety and Health Management,* and the ANSI Z-365 Committee's draft standard *Prevention of Cumulative Trauma*). Although the framework for each of these documents differs, the core elements are essentially the same.

OPTIONS IN SETTING UP YOUR PROGRAM

There are a variety of ways to implement a program, all being dependent on the circumstances of the particular facility. What may work extremely well in one circumstance may not be appropriate at all in others, since individual companies vary in size, complexity of operations, and management style. The level of activity and resources committed to the program depend on the extent of the existing problems. In short, there is no one best way to set up a program. The remainder of this guide provides examples of options in setting up a workplace ergonomics program and further clarification of issues.

Comments:
| Options

Anyone who argues that there is only one approach for implementing any of the basic program elements may be counterproductive in achieving the goal of a safer workplace, since spending time and resources on inappropriate activities can easily distract from making meaningful improvements. Many specifics are simply too new to know for sure what is the most effective approach. Indeed, since all workplaces are not the same, there may not be a "one best way."

ORGANIZATION

The first step is to define objectives and an overall organizational structure. Ideally, a plan should in place before any other activities are started.

Management Responsibility

Responsibility needs to be assigned for overall coordination of the program and for carrying out its various aspects. Authority to make decisions should be defined. Make sure that everyone—including supervisors and employees—knows what is expected of them and understands their role in the process. Good leadership entails clear statements of goals and expectations to all levels of the organization, preferably in ways that are specific and measurable.

Sufficient resources should be committed to address relevant issues. Examples include:

* assignment of staff time
* provision of an ergonomics budget
* retaining an ergonomics consultant

- allowing production employees time off their jobs to become involved in the program
- supporting funding requests for workstation and equipment changes

Accountability should be guaranteed. Managerial reviews should be held to ensure that all parties are following through on their assigned responsibilities.

Comments:
| Management
| Responsibility

OSHA Guidelines refer repeatedly to "management commitment." Unless management's commitment is real and believable, employees will not view ergonomics as important to the business.

The term "commitment" is difficult to define or measure in any particular activity, but often very evident in its overall effect. The employer must be able to demonstrate "commitment."

Two tips are:

Have a "champion" to promote the ergonomics effort. A champion is someone in senior management (who may or may not be in the actual administration of the program) who can be relied on to promote ergonomics within the organization. In large organizations, a champion can be especially important to break through organizational logjams in getting things done.

Be visible. Actions speak louder than words. Plant personnel must be able to *see* that top management is committed.

Exercise—Defining "Commitment"
Pretend you are an OSHA inspector investigating a company with a high rate of CTDs. What would you look for as evidence that the company is making a good faith effort to address their problems?

_____   _____
_____   _____
_____   _____

With this list in mind, what actions would you take in your own company to help show your own management commitment?

Written Policy and Plan

A written plan should identify the organizational structure, outline program goals and objectives, and provide a timetable for achieving them.

The plan should be reviewed and updated periodically to adapt to changing circumstances as well as experiences gained in implementing the program. Objectives for the future should be identified.

Comments:
Written Policy
and Plan

A written plan is not *always* required in order to produce an effective ergonomics effort, but it helps. It is a normal part of business to define a strategy and an operations plan in writing. Implementing an ergonomics program is no different. Writing down the program helps to clarify goals and determine responsibility, routine in any business operation. Many safety professionals consider this step the cornerstone of an effective safety and health program.

In smaller operations, the written plan may be only one or two pages long. In large operations with considerable CTD problems, the plan may take a three-ring binder, including a long section of appendices of the various forms, training materials, and program tools used.

One way to develop this plan is to simply review each item of this chapter and determine the exact approach to be adopted in addressing these items.

Why Bother with a Written Plan?
A written plan serves as the "operator's manual" for your program.

- It helps you clarify goals.
- It helps you sort out assumptions that are shared by some, but not others.
- It forces you to put in black and white many of the things you take for granted—but which might not be understood by others.
- It helps show how a variety of activities fit together as a whole.
- It works to communicate your activities and goals to others; for example, to senior management, new personnel, or regulatory officials.

Ergonomics Team (or Equivalent)

For ergonomic issues, a team approach often helps make for an effective program. Those involved should represent the various plant functions to implement improvements in ergonomics including:

- safety/medical
- engineering
- operations management
- human resource management
- production employees and/or union representatives

- ergonomics professional
- other (for example, maintenance, purchasing)

In some establishments, existing groups such as the safety committee or plant manager's staff already involve the right people and would be natural for this purpose. In this case, there would be no particular need to form a new group.

The purpose of the team is to coordinate the activities described in this chapter, set priorities, make improvements, and document activities.

Comments:
| Team

Once again forming a team is not *always* required. But in many instances, teams can be quite effective for an issue such as ergonomics. Having the multiple perspectives involved can provide good insights on problems and on improvements that will work. Good communications between various parts of the organization are also enhanced.

Employee Involvement

Mechanisms should be established to obtain employee input into the ergonomics process. Production employees often have special insights into ways of improving their own jobs, especially when given training in ergonomic principles. Furthermore, participation often helps pave the way for accepting change.

There are a variety of options from which to choose:

- suggestion systems
- interviews with individuals when job evaluations are made
- formal employee surveys
- department-level Ergonomics Teams that involve employees
- including employee or union representatives in the facilitywide Ergonomics Team
- small group discussions when certain jobs or areas are being addressed
- semiautonomous work groups
-

Comments:
| Employee
| Involvement

Involving employees in the ergonomics effort helps build an effective program. In fact, many people involved in ergonomic design of the workplace strongly believe that employees must be allowed input and given feedback about improving their own work areas. There are simply too many important considerations that people who have never done the jobs cannot be aware of.

Is employee involvement *always* required? Once again, no. Managers or consultants acting independently can make effective decisions, sometimes even more efficiently than when employees are involved. But as a general rule, it is better to involve at least some end-users when you are seeking to make ergonomic improvements in any design. Deliberations may take longer, but the time necessary to implement changes may end up being shorter, since the changes are likely to be (1) thought through more clearly and (2) more readily accepted.

Employee involvement in general has become an effective approach toward improving operations from many perspectives. Many companies have achieved a variety of paybacks because of involving employees in business operations, including improved morale, more informed decision-making, and innovation. A decade or so ago, the concept of employee involvement was unconventional and somewhat radical. Today it appears to be accepted as a good idea, if not a usual way of doing business. Incidentally, if a company has never had a formal way to involve employees, a workplace ergonomics program is an excellent place to start.

As a final comment, in the effort to involve production employees, do not neglect the capabilities of supervisors. This group also often has intimate familiarity with tasks, as well as an understanding of how tasks fit together in a functional, departmentwide whole. Their ideas and input should be sought as well.

Employee-Only Teams?

Some companies have attempted to establish Ergonomics Teams that consist *only* of production employees. Sometimes this approach can be effective, particularly in those locations where semiautonomous work groups are the norm. However, in most facilities this approach can lead to frustration. Managers and others with authority need to be involved in the process, to hear firsthand the observations of the employees and to be able to follow through to implement ideas.

Furthermore, it is often true that employees have limited vision of what is possible. They often accept aspects of the task at hand as a given. In contrast, managers and engineers sometimes have a much better capability for seeing alternative technology for accomplishing a task.

In my view, the interdisciplinary approach works best. Involving multiple perspectives generally produces the most effective process.

Integrate Ergonomics with Other Initiatives

It is a good idea to structure and orient your plan so that the ergonomics process is integrated with other organizational initiatives, such as the quality improvement process or other employee involvement programs. Examples are:

- If your company has invested in developing employee problem-solving teams, use this structure to address ergonomic problems.
- If there is a specific organizational mission statement, show how ergonomics can be used to achieve these goals.

Comments:
| Integrating
| Ergonomics
| with Other
| Initiatives

Careful planning of this sort provides a double payoff: (1) it helps ensure that ergonomics does not appear as a separate or disjointed effort to workplace personnel, and (2) the ergonomics program takes advantage of and contributes to momentum generated by these other efforts. Harmonizing various activities within an organization does not happen by chance, but is the result of foresight and preparation.

TRAINING

Several different types of training are usually needed.

Principles of Ergonomics

Training in basic concepts should be provided to a variety of groups and individuals as appropriate to their jobs:

- management (top, middle, and supervisors)
- engineers (production, process, and maintenance)
- Ergonomics Team
- staff specialists: nurses, human resource, safety, and so forth
- employees exposed to CTD risk factors

All groups should be given an introduction to ergonomics and background pertinent to their roles. For example, managers should be briefed about underlying strategic issues such as the costs of cumulative trauma; the Ergonomics Team should be taught techniques of job analysis and problem-solving; and so forth. Naturally, the depth and extent of this training is dependent on the needs and circumstances of each workplace.

Training sessions also provide an ideal time to solicit ideas for job improvements based on principles of ergonomics.

Comments:
| Basic
| Principles

Because ergonomics is a relatively new issue for most people, special efforts need to be taken to provide some introduction to the topic to almost everyone in a facility. Participation and involvement of all of these groups is crucial, in order to empower them with the knowledge necessary to make improvements plus build ownership and support for change.

Information on Cumulative Trauma

Employees who perform tasks that involve risk factors for cumulative trauma should be provided information on CTDs, the basic symptoms and when and how they should report any problems. Time devoted to such training should be commensurate with the extent of the problem in the work area. Follow-up training may be advisable, such as annual meetings to review basic issues and recent developments.

Comments:
Employee Information on Cumulative Trauma

In many ways, this training is the cumulative trauma equivalent to Hazard Communications training for exposure to chemicals. People who work with hazardous materials have a right to know about both the materials and the hazards. Similarly, people who are exposed to CTD risk factors need to be informed about what to do if they are having a problem and how to help protect themselves. In particular, employees need to be informed that the sooner they report symptoms, the better their chances are for effective (and less costly) treatment. These employee training sessions are also a good time to solicit employee ideas and have a discussion of possible improvements.

Employee Work Methods

Evaluate the need to train employees to use smooth work methods—motions that are rhythmic and not exaggerated or jerky, using minimized repetitions and neutral postures. New approaches to training may be needed:

- Evaluate jobs to find the best work method. Analyzing the differences in work methods between individual employees, or between different shifts or plants can provide valuable information.
- Communicate results between employees.
- Consider assigning personnel to serve as trainers. Traditionally, supervisors have held this responsibility, but many companies are beginning to use special trainers for this purpose.
- Allow adequate time for employee training before the employee is expected to perform at full capacity.
- Consider using special training areas or lines to enable employees to practice or work more slowly until they pick up the work techniques.
- Ensure that employees know how to use their tools and equipment properly.

Comments:
Work Methods

One observation of many ergonomics investigations is the wide range of methods used by employees to accomplish the same task. It is not uncommon to see one employee perform a task smoothly with minimal effort, while the next employee struggles to do the same job. Through experience, the first employee has gained skill and learned techniques that make the work easier and more productive.

Improved approaches to help train employees to use smooth work methods are needed. One new approach is to use various instruments to measure exertion and motion while the employees work. Another useful approach is to videotape employees at work and hold small groups sessions for employees to view themselves at work.

This is not to imply that there is always one best way of conducting a task or that employees should not be allowed variety in how they approach work. Rather, an atmosphere should be created in which employees can experiment with different methods and communicate with each other—always seeking to work smarter, not harder or faster.

Personal Injury-Prevention Strategies

Additional training on personal responsibilities and habits can also be important:

- body mechanics and good lifting practices
- wellness and exercise

Comments:
Personal Injury-Prevention Strategies

Technically, these topics are not part of the field of ergonomics, but they can be important as part of a total CTD prevention program.

Equipment Maintenance

A third area for training, which sometimes needs improvement as part of an ergonomics program, is proper maintenance of equipment. A typical need is for training efforts to be more systematic than in the past, with more follow-up to ensure effectiveness.

Training aspects of equipment maintenance should be reviewed. Primary targets for training are personnel who maintain the equipment, but users may also benefit from this training.

Comments:
Equipment Maintenance

If tools are not used and maintained correctly, extra effort and wasted motions can occur, which merely contributes to cumulative trauma. An example that affects almost every industry is the lack of maintenance of casters and wheels used on carts and dollies.

An often-neglected source of training materials that are directed at specific tools or pieces of equipment are the suppliers themselves. Many suppliers have developed good training materials for the use and repair of their equipment and are willing to provide in-plant training sessions.

COMMUNICATIONS

A good communications plan needs to be established:

With Employees

As the program develops, employees need to be informed on overall objectives and purposes. Progress should be communicated periodically through posters, bulletin boards, employee meetings, newsletters, and the like.

Particular issues to address are:

- Explain why videos and photographs are being taken of jobs, and why questions are being asked about aches and pains.
- Notify affected employees in advance when certain jobs are going to be modified.
- Provide feedback to personnel who have provided ideas for improvement on the status of those suggestions.

Within a Facility

Good communication needs to be maintained between various functions within a company—between operations, engineering, safety, purchasing, and so on. Timely communication is essential to plan properly, coordinate activities, and build momentum for the program.

Between Company Facilities

If a company has more than one facility, mechanisms should be established to share ideas and progress. This can be achieved through informal day-to-day contact, sharing of reports, plus periodic conferences. Video technology provides opportunities to show projects in a way that was impossible in the past.

Comments:
| Communications

Poor communications is a common reason for the failure of many ergonomics programs (not, as is often assumed, the lack of management commitment or the lack of a budget to make changes). Good communications cannot be left to chance, but must be planned, prepared for, and executed. Ongoing feedback is an essential part of empowering people.

If people are not told what is going on, they may assume nothing is happening. For example, changes made in one part of a facility may not be known in other parts. Special steps are needed to keep everyone informed.

Good communications between facilities can help preclude duplication of efforts and promote wise use of scarce resources. Another payoff for these efforts is good cross-fertilization of ideas, which is key for innovation.

JOB ANALYSIS

A good ergonomics program requires the systematic identification of ergonomic issues. In this context there are two basic categories of job analyses: those for documentation and those for making improvements.

- *Documentation*—The goal here is to describe what currently exists; job improvements are not sought. Often, some form of measuring the CTD risk factors is required.
- *Improvements*—Finding improvements often involves two basic stages: (1) a background survey and (2) problem-solving. Measurements may or may not be useful depending on the situation.

Comments:
| Job Analysis

Ergonomics job analysis has been the source of confusion, especially when recommended or required as part of regulatory action. Textbooks and scientific papers often inadvertently give the impression that job analysis is complex and involves elaborate techniques.

The tools of job analysis are varied, some basic and some complex. The technique and approach used must fit the needs and goals of the specific workplace. Unfortunately, too many job analyses are being generated today that are far too long and academic for the need at hand. It is like giving a brain scan to a patient with a head cold, when a simple prescription of aspirin and chicken soup would do.

A good job analysis can be quite simple and yet provide good insight into issues and potential improvements. The key is to be systematic, have a good understanding of ergonomic principles, and be clear about which of several goals you have for the analysis at hand.

Documenting Problems versus Finding Solutions

A primary source of error is not differentiating between documenting a problem versus finding a solution. If the goal is to document problems, then rigor and precision in conducting the analysis is critical. However, if the goal is to find improvements, often the step of determining the problem is fairly simple. The ergonomic issues might be fairly obvious, and there is no particular need to stop to evaluate the problem any further. You can quickly recognize the problems, then almost immediately focus on options for improvements.

This distinction is important, since if the goal is to improve things, then spending time continuing to document the problem can be a distraction and counterproductive. Further study of the problem may lend no insights to problem-solving at all.

Analysis versus Measurements

A second, related source of error is to confuse "analysis" with "measurements." "Analysis" means to separate a whole into its constituent parts. In the context of preventing CTDs, job analysis involves breaking down a job into its elemental steps and identifying CTD risk factors for each separate part of the body (neck, back, wrists, and so on). Measuring risk factors and formally characterizing jobs, while important in its place, may not be necessary. The job analysis can be done using a checklist or indeed without using pen and paper at all.

Good Advice

Early in my career I was given good direction by practical people.

- From a Vice President for Manufacturing: "I don't want a long report telling me that we have problems. I already know we have problems—that's why I called you. Give me a list of things we can do."
- From a Union President: "Don't just tell us all the different ways workers can be hurt. What we need is information on how to make the jobs better."

If your goal is to improve jobs, then focus on solutions, don't just document the problem.

Documentation

The first major category of job analysis involves the precise documentation of job conditions. Usually an expert or highly trained person is needed to do this type of work. Steps for doing these types of analyses are:

1. Determine which jobs are to be evaluated, where to start, and what kind of information is needed. For issues such as the Americans with Disabilities Act (ADA), certain information may be needed for all jobs.
2. Conduct the analyses. Certain measurements are probably necessary. Checklists may be useful, as may videotapes, to be kept as a permanent record. Reviewing and double-checking results are also important, since accuracy is crucial.
3. Update the analyses as needed, since the evaluations need to be changed as the jobs change.

Examples of the need to document job characteristics include:

- *Matching jobs to employee restrictions*—to help determine which tasks are suited to a person with restrictions (either employees with disabilities or injured employees returning to work).
- *Evaluating appropriateness for job rotation*—to help evaluate whether sufficiently different muscle–tendon groups are being used to prevent CTDs.
- *Proving whether ergonomic issues are present or not*—either to management or in reference to regulatory or legal matters.
- *Conducting "before-and-after" studies*—to evaluate the effectiveness of job improvements.
- *Advanced Task Analysis*—to be more precise when common sense and simple evaluations are not sufficient.
- *Conducting epidemiological studies*—to compare job conditions with health effects.
- *Documenting background information*—to keep a record of background data on each job.

Improvements

The other major category of job analysis is finding ways to improve jobs. Finding improvements often involves two stages: (1) a background survey and (2) problem-solving.

Background Survey

The goal of this first-stage survey is to answer these questions: Is there a problem in our workplace? Where are the problem areas? Which problems do we tackle first?

Injury/Illness Records

As part of a safety program, the usual first step is a review of injury/illness records and workers' compensation data. Several reviews are typically made, for both cases and rates (cases per exposed employees):

- OSHA-recordables
- lost time
- workers' compensation costs

Comments:
Injury/Illness
Records

It is not unusual for different jobs to appear the worst, depending on the criterion used. These "conflicting" results are not a source of concern. Each is important and reviewing all helps give a complete picture.

Analyzing workers' compensation data has the added benefit of providing *costs* of disorders per job (or department).

Error/Defect Analysis

Similarly, other background evaluations can be conducted to identify problem areas. As part of a quality improvement process, errors and defects can be reviewed with an eye toward relating these issues to "human" error. These human problems can perhaps then be corrected through ergonomic improvements.

Comments:
Error/Defect
Analysis

Often it is valuable to review these issues, since those jobs that are causing problems for people often are problems for quality and production as well.

Employee Surveys

Quite often, available recorded data are not complete. Injury/illness records are dependent on employee awareness—or lack thereof—to report problems. Additional ways to solicit information from employees may be needed, including:

- *Informal discussions*—An effective and informal approach is to simply ask people where they have experienced problems. Issues can be discussed either in group meetings or individually on the job.
- *Questionnaires*—Conducting a survey using a self-administered questionnaire is also a useful technique. For many ergonomic issues, an employee Discomfort Survey (or Symptoms Survey) is helpful. These are usually simple one- or two-page questionnaires that ask employees if they experience physical discomfort from their jobs, usually by referring to a part of the body and an index of severity. Typically, these are anonymous, identifying only the department (and task, as appropriate).
- *Measurements*—Additionally, new, more objective techniques are being developed that can be used in a survey fashion. Such approaches include vibrometry or various neurometry instruments for measuring nerve entrapment disorders such as carpal tunnel syndrome.

Comments:
| Employee
| Surveys

Simply talking to people about where they have experienced problems is the simplest and probably the most effective approach. Both supervisors and production employees often have a good intuitive understanding of which jobs need improvement and what the specific issues are.

The advantage of the questionnaire method is that good questionnaires permit results to be graphed or statistically analyzed. This allows comparisons between jobs or departments, and "before-and-after" studies. Often the results of these types of proactive surveys are more accurate and sensitive than an OSHA 200 Log analysis, which depends on employee awareness to have reported problems.

Use of a measurement instrument such as a vibrometer (a device that measures the ability to feel sensations in the fingertips) for this purpose is still in an exploratory stage, but the concept has been used successfully in pilot studies.

Personnel Data

Often a review of various types of personnel data—either by studying records or by talking with supervisors and employees—can help identify problem jobs. Examples are jobs that:

- have high turnover
- are universally disliked
- are lowest on bid lists
- are used as entry-level because of undesirability

Comments:
| Personnel
| Data

These undesirable jobs may be prime targets for ergonomic improvement. Sometimes they do not show up on an OSHA 200 Log analysis because people do not stay on these jobs long enough for cumulative disorders to occur. In fact, they may spuriously affect the injury rate of other jobs.

Walk-through Survey

Another approach to background identification of problem areas is to briefly review each job for ergonomic issues. Often, recognition of some problem areas emerges only by seeing the task.

Comments:
| Walk-through
| Survey

It is helpful to look for ergonomic "triggers" such as awkward postures, employees wincing when they do certain actions, or makeshift changes, such as with masking tape or cardboard. A simple one-page checklist may also help.

Problem-Solving

Based on the results of the background surveys, decisions can be made on priorities. Then attention can be focused on these critical areas, one job or task at a time, using problem-solving techniques:

- Identify the steps of the job.
- Identify the specific risk factors or ergonomic issues for each step. It is not necessary to measure the risk factors, although rough estimates of "low," "medium," or "high" for each affected part of the body can be useful.
- Identify options for improvements using a variety of techniques:
  - common sense
  - brainstorming
  - root-cause analysis
  - cause-and-effect (fishbone) diagrams
  - process flowcharts
  - process control charts
  - other

PROCESS

A very useful approach for problem-solving involves a team approach:

- Fact-finding with checklist by two or three team members
- Discuss issues with employees
- Videotape the job
- Review video and checklist results with whole team
- Brainstorm options for improvement
- Plan action
- Evaluate results

Comments:
| Problem-Solving   Well-designed checklists and work sheets can serve as simple, yet effective tools. Supervisors and employee representatives can easily be trained to identify basic issues using a good checklist with virtually the same accuracy as a professional ergonomist. A good checklist can serve as a "mind-jogger" of risk factors and can help the evaluator to be systematic and to break the task down into specific elements.

The problem-solving tools of Total Quality Management are

often effective techniques. Reviewing videos is also an excellent technique. Videos help focus attention—typically, you will notice issues on the video that you do not see live in the workplace. Finally, the team approach to identifying issues and options for improvement has proved to be quite effective in the workplace.

A Final Note on Quantification

While keeping in mind my previous warnings about confusing your goals, sometimes it is necessary to measure the risk factors in order to solve problems (for example: How many repetitions per shift are there? How much force is involved?) Examples of when quantification is necessary are:

- When common sense and the simple evaluations described above are not sufficient:
  - identifying relative stresses and risks among different tasks
  - identifying peak stresses in a single task
  - communicating to others the results of your analysis
- When alternative choices for potential improvements must be evaluated.
- When "before-and-after" studies are desired.

MAKING JOB IMPROVEMENTS

Finding ways to improve jobs is the key part of an ergonomics program. There can be countless options for making improvements, all based on the basic principles of ergonomics. The only limitation arises from inhibited thinking.

Short-Range Improvements

After initial job evaluations are made, many small and inexpensive improvements can often be identified. Sometimes fixes can be completed literally overnight and other times over a period of months. Typical examples are layout changes to improve heights and reaches, the purchase of antifatigue mats and chairs, or changing work methods.

Long-Range Equipment Development

Long-range improvements are sometimes also needed. This may involve some research and development before a change becomes technically feasible. In many cases, these changes may dovetail with long-range plans for quality and productivity improvement.

Comments:
| Improvements

Whether short or long range, everyone should recognize that ergonomic improvements are not always straightforward and that a period of experimentation and trial and error is often needed to find a good modification. The concept of "continuous improvement" is important—a job improvement is planned, then implemented, then evaluated, and then refined in an ongoing process.

Sources of Ideas

In-House Capabilities

Once the basic concepts of ergonomics are understood and problems identified, almost anyone can have an idea for improvement. Your in-house engineers, managers, and production employees can be the source of many improvements once they are pointed in the right direction.

Literature

An increasing amount of information is becoming available on improvements:

- industry trade magazines, including magazines from industries *other* than your own
- publications on ergonomics
- the general ideas included in the OSHA guidelines
- vendor catalogs—a wealth of ideas can be found in these catalogs, especially if you are creative and can think of unconventional uses of standard equipment

Comments:
| Remember:

"Ergonomic" products that are used inappropriately are not ergonomic.
Any product that leads to improvements can be "ergonomic."
There is no certification or testing procedure to determine if a product is truly ergonomic—buyer beware!

Ergonomics Consultants

Ergonomists can be a useful part of this process, by challenging traditional thinking and by providing new insights into the design concepts. The ergonomist can provide a new set of eyes to see workplace issues to which you may have become accustomed. Finally the ergonomist may have experience with improvements in other industries that may be of value to you.

Contractors and Equipment Suppliers

A good idea is to invite contractors and equipment suppliers to your meetings and training sessions. Many companies use local contractors to build and install equipment. Make sure that they know of your efforts and that you expect equipment that has undergone ergonomic review.

Other

Additional sources of ideas can be found by visiting other facilities, both in your industry and in other industries. Conferences and trade shows can also be valuable.

Tracking System

As part of this process, a good filing system may be needed to keep track of results of job evaluations, ideas for improvement, planned changes, and overall progress. The system can range from a simple notebook or file drawer to a computer spreadsheet or data base system.

MEDICAL MANAGEMENT

Medical management refers to the full range of activities that health care professionals do to address CTDs—assessment, diagnosis, treatment, medical restrictions, and proper return-to-work of employees with CTDs. The development of a standardized, comprehensive medical management component integrated in the ergonomics effort is critical to the success of the overall program.

Although the medical detail on each of these developments is outside the focus of this book, it is important for managers to recognize that medical management of CTDs should be part of an overall program. Furthermore, from a manager's point of view it is important to ensure that certain elements are in place. A number of steps can be taken to minimize the effects of CTDs on a working population:

Early Recognition

Systems should be established to ensure that people with symptoms are identified as early as possible. Methods for accomplishing this goal are beginning to emerge and include:

- training sessions in which employees are instructed to report problems

- providing periodic physical exams to employees whose jobs involve cumulative trauma risk factors (referred to as "active medical surveillance")

Systematic Evaluation and Referral

Steps should also be taken to ensure that (1) standardized diagnostic tests are given when employees report symptoms and (2) standardized procedures are followed for treatment and referrals. Protocols to provide guidance in this regard have been developed in recent years.

Conservative Treatment and Follow-up

The primary goal is to treat any health problem at an early stage and avoid surgery or other more radical treatments. Furthermore, mechanisms should be established to ensure that employees return to work only when ready, are put on jobs that are compatible with any restrictions, and are evaluated periodically to see that problems are not recurring.

TERMINOLOGY

Ergonomics versus *CTD Prevention*

It is important to remember that *ergonomics* and *CTD prevention* are not synonymous. There is a strong overlap, but technically, medical management is not a part of ergonomics.

Case Management versus *Medical Management*

Some experts distinguish between *case management* and *medical management.* The former refers to something done by nonmedical personnel, while the latter is an area for professional medical providers. In essence, *case management* means knowing about the circumstances of each case and monitoring the status of the individual at each stage of the treatment process to ensure that the person does not get lost in the system. *Medical management* means diagnosis and treatment of people from a strictly medical point of view.

In this book, a broad definition of *medical management* is used to entail all of the case management strategies and practices that go along with administering the medical aspects of preventing CTDs.

Integration of Health Care with Ergonomics

Active steps should be taken to ensure that health care providers are fully integrated with the workplace ergonomics program. Examples of such steps include:

- inviting health care providers to be members of the ergonomics committee
- providing health care providers with tours of the workplace
- communicating workplace activities with health care providers

Comments:
| **Medical**
| **Management**

The medical management of employees' cumulative trauma has been one of controversy and disagreement. Both evaluation and treatment methodology of CTDs are rapidly evolving. As more experience is gained, consensus will probably develop. What *is* important is that employers take proactive steps to begin to address the issues.

In the past, the treatment of choice for CTDs like carpal tunnel syndrome was surgery, expensive for employer and not always in the best interest of the employee. There is now much greater emphasis on conservative treatment. The goal is to identify the disorders early when treatment options are better. If identification is made early, the condition can be treated with a variety of simple measures that reduce swelling and allow natural healing: rest, anti-inflammatory medication (aspirin, ibuprofen), ice packs, and rehabilitation therapies. By waiting too long, expensive surgery may be needed, with lower chances of complete recovery.

| **How CTD**
| **Medical Issues**
| **Are Different**

Most guidelines or standards for occupational disorders contain sections on medical management, such as the NIOSH Criteria Documents or OSHA standards for various health hazards. Among other issues, the medical management programs address surveillance of employees and, in some cases, treatment. Medical management of cumulative trauma is no different in concept from these other health hazards.

However, there is a basic difference between cumulative and other occupational hazards that complicates matters. This difference can perhaps be best described by contrasting cumulative trauma with hearing loss.

The diagnostic test for hearing loss, the audiometric examination, is (1) well accepted and fairly objective. But (2), unfortunately, once hearing is lost because of noise, there is not much that can be done for treatment—hearing loss is permanent.

With cumulative trauma, it is exactly the reverse:

1. *Evaluation can be difficult.* There is little consensus concerning diagnostic tests for cumulative trauma. The tests themselves differ substantially depending on which part of the anatomy is affected. Furthermore, the tests are not as objective

as diagnostic techniques for many other diseases. One of the most common symptoms of cumulative trauma is simple pain, which cannot yet be measured or determined independently of employee statements or complaints.

2. *But CTDs are treatable.* If the CTD can be identified sufficiently early, there is a great deal that can be done for treatment.

Thus, from a management perspective, the medical issues become exceedingly important. Medical evaluation programs are complicated and take time to develop and implement. But the effort is worthwhile.

Medical management programs for CTDs involve the development of a special system to handle the ambiguities and the varieties of medical problems. The questions are:

- Surveillance: How do we find out who has problems?
- Evaluation: By what criteria are people evaluated? And for which disorders—lower back, shoulders, elbows, wrists?
- Referral: When can ailments be treated in the workplace? And when should people be referred to specialists? By what criteria are these decisions made?
- Treatment: What treatments are effective? In particular, what treatments are effective in the early stages? When should cold be applied, and when should heat? Or when should the patient exercise and when not? And what type of exercise? And for what part of the anatomy?
- Return to Work: When is the employee able to return to work? Can the employee return to normal work? Or is restricted duty in order? What kinds of tasks are appropriate for restricted duty? For how long? How do we monitor employees to ensure that they are not being reinjured from their work?

It is because of these issues that the medical management of cumulative trauma has surfaced as a controversial issue. It is inattention to all of these questions that leads to treatment failure and thereby raises disability costs for employees. The OSHA Guidelines for Meatpacking have provided in many ways the first comprehensive overview of proper procedures. As a start, these guidelines may be useful, especially if they are seen as examples of approaches, rather than precise requirements.

| **Workplace Programs**

The need for early recognition heightens the importance of good communications with employees. Effective medical management of an ergonomics program requires knowing as soon as possible when employees are hurting. Recorded rates may increase but costs per case should drop drastically.

To maximize the value of a good medical program, there must be good communication and involvement between the health care providers and the coordinators of the workplace ergonomics program. Some occupational health physicians believe that medical providers who do not involve themselves in the workplace are not qualified to treat CTDs.

PREDICTING WHO WILL GET A CTD

Efforts to develop practical tests that predict who is likely to develop a CTD have been unsuccessful. Even if there were ways to distinguish between people prior to employment or assignment to a task, there are legal and ethical questions associated with this approach, especially now in light of the ADA. Furthermore, efforts to develop such predictive tests shortcut the possibilities for making innovations in the workplace that increase productivity and quality.

Thus, screening programs to predict who will have problems are not generally recommended as part of a medical management program.

MONITORING PROGRESS

The ergonomics program should be evaluated periodically. Several approaches can be used.

Management Review

Overall activities should be reviewed by top management on a regular basis to ensure that goals and objectives are met and to modify the program accordingly.

Comments   Top management review of the program usually is crucial to the success of a program.

Injury/Illness Trends

Cumulative trauma rates for the specific areas of their occurrence should be calculated on a regular basis to identify any trends.

Comments   Monitoring these trends is essential. Graphs work well for making comparisons and showing these trends.

Ergonomics Log

It is helpful to maintain a log of all job improvements made that have an ergonomic impact. This log may include items such as: department, job, ergonomic issue involved, improvement, cost, date, contact person, and so forth.

Comments   Keeping this log provides value in the following ways:

- The log reminds you of your accomplishments, thus keeping momentum going and the morale of the Ergonomics Committee high. People tend to forget all of the things that we do, and if we do not keep a running list, we feel that we have not accomplished much and our morale suffers.
- The log provides a basis for communicating results to both managers and employees. It can be impressive to report, as some companies have, that "in the past year at the XYZ Company we have made 326 improvements in the workplace as a result of our ergonomics effort."
- The log can provide evidence to regulatory agencies of your commitment to ergonomics.

Special Studies

A variety of issues can be evaluated to measure the effect of the ergonomics program:

- job analyses before and after improvements are made
- employee survey results
- workers' compensation costs
- turnover and absenteeism
- quality and productivity

Comments   The approach taken here depends greatly on the specifics of each workplace. In some cases, no special studies are needed—simply monitoring the injury trends is sufficient. In other cases, being able to show successes and benefits takes considerable extra effort.

MORE DEBATES—EXERCISE AND SPLINTS

The following are a few questions that typically arise with regard to prevention of CTDs. Technically speaking, none of these are really part of the field of ergonomics, but are often asked of ergonomists.

Does Exercise Help Prevent CTDs?

Exercise and conditioning programs may have merit as part of a CTD prevention program. Everyone from the CEO to the most recent new-hire should be in shape, warm up, and stretch during the day. The established fact that athletes must train, warm up, and stretch to avoid injury gives credibility to the concept of the "industrial athlete." Definitive proof, however, is not available; the studies that have been done show mixed results. It may well be that the question "Does exercise help prevent CTDs?" is too broad in scope. The real question may be "Which types of exercises help prevent what types of CTDs, for which tasks and under what conditions?"

For example, it has become increasingly clear that a strong and flexible body helps prevent cumulative back injuries, but the impact on upper extremity disorders is highly debatable. Furthermore, the effect may be different depending on whether the task is physically demanding versus light and static.

In any attempt to resolve this controversy, it is helpful to start breaking the topic down into specifics, listing both the pros and the cons.

The Pros

There are several kinds of exercise, each having its own purpose:

1. *Warm-up*—One type of exercise is the warm-up at the beginning of the shift and before returning to work after lunch. This type of exercise is reminiscent of Japanese work culture, which begins the day with the company song and some exercises. U.S. companies are also beginning to implement this type of exercise program as part of an effort to reduce CTDs by permitting employees to limber up muscles before conducting strenuous work. The sports analogy is very relevant here. Athletes who warm up properly suffer fewer injuries; an equivalent experience is emerging from industry. This type of exercise should be good for everyone.

2. *Regular stretch breaks*—A closely related type of exercise is a routine stretch break taken throughout the day. These breaks serve to stretch unused muscles and provide relief from being in the same position for too long. People feel more refreshed and are better able to do their jobs comfortably. This type of exercise would be best for those with static jobs.

3. *Conditioning*—A third approach is a true conditioning program. Some companies are moving beyond mere stretching and warm-up programs into heavier exercise, with a goal to get people in shape by strengthening muscles and improving flexibility and endurance. There is logic to the argument that being in shape reduces the risk of cumulative trauma. Simply being stronger can reduce the exertion needed to perform a task, in the same manner that an eight-cylinder car can go up a hill easier than can one with four cylinders. Good flexibility may also play a role. Furthermore, the increased endurance that comes with being in shape helps prevent fatigue, which often is a precursor to injury. And finally, there is evidence that being in good condition speeds recovery time following an injury. Conditioning programs probably would work best for physically demanding jobs.

4. *Physical therapy*—A final type of activity is physical therapy. Here, certain exercises are used to help treat musculoskeletal disorders. Considerable experience and research in the field of physical therapy points to the useful role of exercise in treatment.

All four of these approaches have merit for meeting each separate goal. Much of the controversy that surrounds exercise in the workplace can be resolved by distinguishing between these types and goals of exercise.

The Cons

Exercise may be good for everyone, but the degree to which exercise can prevent CTDs may not be as straightforward. To help explain, the athletic analogy is again useful. Most people could not run a marathon, for example, without a conditioning period to get in shape (and certain jobs have requirements that are analogous to running a marathon every day). And anyone who is about to run a marathon would certainly warm up prior to starting the race. However, it is also true that many marathon runners suffer from cumulative trauma of the legs, no matter how good shape they are in and how much they warm up prior to competition. The human body simply is not designed for this kind of repetitive and forceful activity, especially if done day after day.

Furthermore, simply being stronger may not be relevant at all. For example, one of the most important issues for back injuries is the load placed on the discs in the spinal column. Stronger muscles in the back may not reduce the actual load on the discs themselves at all. On the contrary, the load on the disc may increase—a stronger person may weigh more than a weaker person, plus the stronger person may be tempted to lift heavier objects or ignore good techniques.

Another aspect of the issue is the type of exercise prescribed. For example, in some facilities employees have been instructed to squeeze tennis balls to strengthen their forearms in an effort to help prevent wrist disorders, despite the fact that their jobs required similar motions. These exercises might have been exactly the wrong thing to do. Stretching the forearms in a pattern opposite to that of the task might be more effective.

In short, being in good physical shape may help, but it is not the total answer. In the workplace, exercise programs should not be undertaken in lieu of addressing the demands of the jobs and making improvements where feasible, but, rather, should be used to augment their preventive and productive impact. As argued throughout this book, ergonomics serves as a tool for management to identify and achieve workplace innovations. Relying strictly on exercise as a way to prevent CTDs robs management of this tool, circumventing the possibility of finding better ways of doing tasks.

Incidentally, it is useful to remember that exercise per se is not ergonomics (exercise does not change a task or tool to make it fit people more compatibly). This might be splitting hairs, but it does help explain and define the boundaries of ergonomics.

Are Musculoskeletal Supports Helpful?

In the past few years, there has been considerable debate about the use of back belts and wrist splints for injury prevention. The main points of the debate on this issue are outlined below.

Medical issues concerning back belts and wrist splints will ultimately be resolved. But from a purely business point of view there are still two points to keep in mind about these supports:

1. The most important shortcoming of back belts and wrist supports is that, by relying on these devices, a business gives up the possibility of making productivity and quality improvements. As long as a company is attempting to make workstation and equipment changes, there is hope that some breakthrough in innovation can be achieved. By relying on musculoskeletal supports a business gives up all hope of finding a better, more competitive way of doing the job. The solution to back injuries among ditchdiggers was the invention of highly productive backhoes and draglines, not distributing back belts.

2. In general, these supports should not be used in lieu of a complete prevention program. It would be a shortsighted decision to purchase these supports by themselves and not address the fuller issues.

Most ergonomists have a bias against the use of back belts and wrist splints, since these devices are the antithesis of ergonomics—they are intended to help people adapt to tasks and do nothing to change a task to meet human limitations. However, as part of a total CTD prevention program, there can be a use for musculoskeletal supports, either as part of treatment or, in the case of back belts, when engineering-type improvements have proven to be infeasible.

Back Belts

Back belts are designed to reduce spinal compression forces during the lifting of heavy objects. These belts are thought to work much in the same way that tightening the abdominal muscles helps to shift part of the load from the spinal column to the intraabdominal area. There seems to be high employee acceptance of the devices and, if nothing else, back belts appear to provide a benefit by helping remind employees to use good body mechanics while lifting.

There is a negative view challenging whether these supports provide any meaningful physiological help. There is also some concern that the supports have side effects such as contributing to a weakening of muscles and ultimately leading to more injury as a result of dependence on them rather than good muscle tone.

The results of studies on whether back belts actually reduced injuries are mixed: some show a positive benefit, while others show either no benefit or a negative effect (in particular, people who wear back belts and then stop the practice may suffer higher injury rates than normal). The discrepancies may be explained by the industry being studied (they were done in different industries), the type of study conducted, and the type of belt used.

Wrist Splints

Wrist splints are designed either to immobilize the wrist or to provide some rigidity to support the wrist. Their primary use is as part of treatment, such as by wearing them at night to ensure that the wrists maintain a neutral position during sleep.

However, if worn routinely at work in an attempt to prevent disorders from occurring, wrist splints can make matters worse. There have been cases where companies that required employees to wear wrist splints found that CTDs actually increased. The problem is that if the task requires the employee to bend the wrist, the splints can force either of two negative side effects:

* The repetitive motions are simply transferred to another area, primarily the shoulder. The wrist is kept straight, but this is compensated for by awkward elbow posture and increased shoulder motion.
* More exertion is required. By wearing a splint, employees are in effect required not only to continue to do the damaging task, but to bend the wrist support as well. The splints thus increase the effort involved in a wrist motion, which multiples the risk of getting a disorder.

Some wrist splints are more flexible than others. Nonetheless, these devices still have the potential of increasing the risk for CTDs. Thus, if used at all, wrist splints should be used only sparingly as part of treatment and not part of any prevention effort.

INDUSTRY GROUPS

Some problems are beyond the control of any individual company and can only be dealt with on an industrywide basis. Consequently, both trade associations and unions have a valuable role to play, particularly in developing information on problems and solutions in specific industries and encouraging cooperation within an industry to help address ergonomic issues. Other issues, such as creating training materials or conducting research, can be done company by company, but usually are more cost-effective if done through an industry group.

Two examples of positive action based on my experience are the American Meat Institute (AMI) and the United Auto Workers (UAW):

American Meat Institute Program

Over the past several years, the AMI has developed a comprehensive program in ergonomics to aid its members. Few, if any, other industries can demonstrate an equivalent program involving as coordinated an effort on an industrywide basis.

Organization

- established AMI Safety Committee composed of member safety directors to coordinate activities
- established AMI Human Factors Committee composed of member vice presidents of engineering to help accelerate the development of improved equipment
- involved equipment suppliers in programs
- retained an ergonomics consulting firm for ongoing assistance
- established an awards program administered by the National Safety Council to promote program development at a plant level

Training

- developed training materials:
    - managers' guide to ergonomics
    - plant guide to setting up an ergonomics program
    - employee handout on ergonomics
    - videotape for employees
- held numerous regional and national conferences on safety and ergonomics
- provided presentations and discussions of ergonomic issues at regular AMI meetings: for example, management conference, human resource conference, legal committee meetings

Communication

- held regular meetings to improve sharing of information within the industry
- established a Safety/Ergonomics newsletter to keep members informed on developments

- initiated contact with foreign associations, agencies, and scientists to learn of developments in meat operations elsewhere
- cooperated with the trade press to improve sharing of ideas within the industry

Research/Identification of Risk Factors

- funded pioneering studies on the use of vibrometry as an ergonomics survey technique
- funded a study on optimal working postures through the University of Nebraska
- initiated a U.S. Bureau of Labor Statistics study on CTD rates by workplace size
- funded expert ergonomics assessments, shared with the industry
- met on several occasions with NIOSH scientists to discuss long-term research needs
- sponsored a meatpacking section meeting at the 1989 Human Factors Society Meeting
- described ergonomic innovations at the 1990 National Safety Congress

Identification of Improvements

- funded a study on potential applications of new technology
- sponsored a safety pavilion at the AMI trade show to promote awareness of new equipment with ergonomic implications
- monitored European equipment developments
- promoted cooperative R&D efforts among member firms

Medical Management

- conducted education on advancements in medical management of CTDs
- evaluated medical management recommendations provided by government agencies and others

These efforts have proven successful by helping the industry reduce the rates of CTDs.

United Auto Workers

The UAW has maintained a large, professional health and safety staff for decades. The resources that the UAW has invested in health and safety in general, and ergonomics in particular, are impressive, and actually superior to those of many companies. The union has played a unique role in promoting industrywide efforts to combat problems, not only by bringing issues to management's attention, but also by helping to solve problems themselves. As examples of activities related to ergonomics, the UAW has:

- met regularly with companies in a nonadversarial atmosphere to discuss joint problem-solving
- promoted research
- developed training materials
- conducted industry-specific training sessions
- helped develop and find solutions

ADAPTING FOR ORGANIZATIONAL CHANGE—PRACTICAL APPLICATIONS

- *Provide a vision* of where you want to go and what the future organization will look like. This should be concise and easily understood by everyone. It often involves a statement of what is possible, presenting means to overcome barriers imposed by the limitations of the past. Create a single image or metaphor to help people grasp this overall glimpse of the future.
- *Develop a coordinated, systematic plan.* Details of the change need to be worked out along all dimensions of the organization: facilities, process technology, human resource policies, and so forth.
- *Use a team.* Developing and implementing the plan should be a team effort, involving representatives of all functions of the organization.
- *Involve people in the change.* Participation creates "ownership" of the new ideas and helps people to buy into the process, rather than having it dictated from above. Involvement also helps provide good ideas to improve design.
- *Provide information* on (1) the need for change and (2) the skills and roles that will be essential for the change.
- *Provide "testimonials"* to establish the need for change, for example, "I have seen the errors of my ways, and am now setting forth on a new road. . . . "
- *Appeal to self-interest,* using the framework and language of your audience. For upper management, talk about costs and benefits. For supervisors, explain how the change will make their lives better, for example, by reducing administrative work load by fixing the cause of problems. For employees, explain how the changes will make work easier. (Employees are usually an easy sell on the need for change. The more typical need is to show management commitment, which is done best by action, not words.)
- *Appeal to higher instincts.* Do not neglect to refer to how the change is "the right thing to do" and how it can benefit people or society as a whole. It can be surprising how often people want to do the right thing for the right reasons.

- *Describe your message in familiar terms.* Subjects like ergonomics can sound difficult and foreign, so try to explain the program in terms your audience identifies with. Do your best to demystify the subject, showing practical examples.
- *Think in terms of converts.* In trying to change an organization, you may need to gain support one person at a time.
- *Reach a critical mass of people.* Once you have converted a sufficiently large number of people, you will attain a critical mass. Afterward, change can become much easier.
- *Establish a demonstration project.* If an entire system cannot be affected all at once (which is often the case), change just part of it to provide a model for everyone else. It also gives you a chance to work out the bugs in your program at the same time.
- *Reinforce the message from all directions.* Try to structure the setting so that wherever people turn, they receive the same message. The quality message has crept into virtually all aspects of our society in the past decade. Information on ergonomics is now emerging, such as in advertisements.
- *Use "organizational judo."* Take a tip from martial artists and take advantage of the momentum of your organization. You may not need to take everything head-on—all you may need to do is flip your organization in the right direction.
- *Improve continuously.* One of the tenets of quality is that there is really no end to the change process. Improvements must continuously be made. No one should have the expectation that an ergonomics program will be put in place and at some point be over and done with. Continuous improvement is the name of the game.

BEHAVIORAL CHANGES AND ERGONOMICS

The following guide contains helpful ways to get people to use new methods and equipment.

- *Make sure the change is better.* Particularly with work methods, sometimes we think we know the right way, but there are often other issues that confound our thinking. The ergonomic principles themselves may not be sufficiently precise to be of help in designing many tasks. There may be subtle issues in the work task that we do not understand. Try to understand all of the reasons why it may not be so simple to change a behavior.
- *Make sure the change is easy to use.* The classic example is installing hoists to eliminate the need to lift. But hoists are often slow and clumsy and people resist using them. Conveyors, skids, and slides are often used more readily.
- *Train, involve, and empower.* Explain what, why, and how. For example, with

adjustable office chairs, people need to be told how the chairs are adjusted and the benefits they can provide.

- *Provide practice.* Instructions are not enough. People must practice a technique for it to become habit.
- *Provide positive reinforcement.* People respond well to praise.
- *Try it, you'll like it.* Sometimes, people have to simply be talked into trying the new design for a while. The length of trial must be long enough to overcome old habits.
- *Disrupt everything (sometimes).* On occasion, people are so ingrained in old habits and have such strong muscle memory for certain work methods that it is virtually impossible to make changes, especially minor ones. Sometimes it is easier to disrupt the works, so that everything is new and the task has to be learned again from scratch.
- *Modify reward systems.* If people are working under incentive systems, make sure they are not penalized for taking the time to try new methods. Sometimes the new method may involve working slowly for a while to learn how to do it correctly.

Don't Do Work the Hard Way

Q.   Does it make any business sense to require either employees or customers to do unfriendly tasks such as the following?

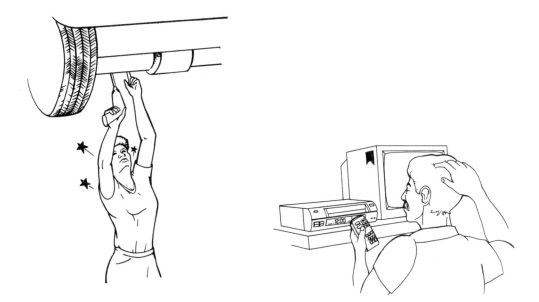

Physical—Working with your arms overhead while looking upward for a period of time.

Cognitive—Trying to understand how to operate a piece of equipment.

A.   Yes _____   No _____

I have tried to make the case that ergonomics ought to be a part of every business strategy and that it can provide a competitive edge to any organization. I have shown by explanation and example that it is a broad field that provides new tools for management problem-solving in a multitude of ways. I have described how the field furnishes concepts and techniques that can improve efficiencies of the workplace and heighten the appeal of products.

Ergonomics is becoming part of the business world if for no other reason than it is being required as part of regulatory action. But it does not follow that it is just another burden on commerce, that is, if you do it right and are creative in how you think about the issues relevant to your situation. It doesn't mean just providing "cushy" jobs—it means being more efficient.

Ergonomics can be a complex field, and in its details and advanced techniques (which were not presented in this book), it can be a demanding and rigorous discipline. But in practical application it does not need to be, especially in initial phases on any given problem. Much of ergonomics can be common sense. The most important objective for any organization is to grasp the basic principles and apply them systematically.

I have tried to show that the underlying thrust of ergonomics is a tradition, rather than an untried theory. In many ways, the core of ergonomics involves human intuition, rather than some esoteric philosophy.

Currently the tools of ergonomics can provide value to any business, but the next decade will provide even greater benefit, including:

* precise guidelines for product and equipment design
* better ways to measure
* further proof of its value

To me, the greatest satisfaction of working in this field has been the fresh insights it provides for problems—putting on our ergonomic glasses and seeing things from a new perspective. Ergonomics furnishes a vital stimulus for innovation, inventiveness, and simply finding better ways to get things done.

REFERENCES

CHAPTER 1

American Productivity Center. 1982. *White Collar Productivity: The National Challenge*. Grand Rapids, MI: Steelcase, Inc.

Brown, R. Todd, George Page, and Paul McMahan. 1991. Ergonomics in the U.S. railroad industry. *Human Factors Bulletin* 34(3).

Dainoff, Marvin. 1990. Ergonomic improvements in VDT workstations: Health and performance effects. In *Promoting Health and Productivity in the Computerized Office,* eds. Steven L. Sauter, Marvin J. Dainoff, and Michael J. Smith. London: Taylor & Francis.

Francis, Joellen, and David L. Dressel. 1990. Workspace influence on worker performance and satisfaction: An experimental field study. In *Promoting Health and Productivity in the Computerized Office,* eds. Steven L. Sauter, Marvin J. Dainoff, and Michael J. Smith. London: Taylor & Francis.

Gilbert, R., R. Carrier, F. Laurent, and J. Schiettekatte. 1990. Improving the work area: A case study. *Ergonomics* 33(3):375–381.

Harris, Douglas H. 1987. *Human Factors Success Stories.* (video) Santa Monica, CA: Human Factors Society.

Noro, Kageyu, and Andrew Imada. 1991. *Participatory Ergonomics*. London: Taylor & Francis.

Oxenburgh, Maurice. 1991. *Improving Productivity and Profit Through Health & Safety*. Sydney, Australia: CCH International.

Rawling, Richard, and Paul O'Halloran. 1988. Effective control of manual handling injury in the electricity industry. In *Proceedings of the 10th International Congress of International Ergonomics Association,* Sydney, Australia, pp. 266–268.

Schneider, Franz, and Dennis Mitchell. 1989. Look up, look way up. *OH&S Canada* 5(2):87–88.

Spilling, Svein, Jakob Eitrheim, and Arne Aarås. 1986. Cost-benefit analysis of work environment investment at STK's telephone plant at Kongsvinger. In *The Ergonomics of Working Posture,* eds. N. Corlett, J. Wilson, and I. Manenica. London: Taylor & Francis.

Springer, T. J. 1986. *Improving Productivity in the Workplace: Reports from the Field*. St. Charles, IL: Springer Associates, Inc.

Steele, S., R. Hamel, J. Muller, and J. L. Wick. 1990. Wrist injury prevention in firearms manufacture: A case study. In *Advances in Industrial Ergonomics and Safety II,* ed. Biman Das. London: Taylor & Francis.

Sullivan, Carol. 1990. Employee comfort, satisfaction and productivity: Recent efforts at Aetna. In *Promoting Health and Productivity in the Computerized Office,* eds. Steven L. Sauter, Marvin J, Dainoff, and Michael J. Smith. London: Taylor & Francis.

Thomas, Robert E., Jerome Congelton, and William H. Horsford. 1989. An ergonomic evaluation of customer service operations—A case study. Paper read at Human Factors Society 33rd Annual Meeting, 16–20 October, Denver, CO.

Thompson, David A. 1990. Effect of exercise breaks on musculoskeletal strain among data-entry operators: A case study. In *Promoting Health and Productivity in the Computerized Office,* eds. Steven L. Sauter, Marvin J. Dainoff, and Michael J. Smith. London: Taylor & Francis.

Webb, Robert. 1989. A feeding frenzy. *OH&S Canada* 5(2):91–92.

Westinghouse Architectural Systems Division. 1982. *A Study of White Collar Productivity and Open Office System Furnishings: An Executive Summary.* Grand Rapids, MI: Westinghouse Electric Corporation.

Wick, J. L., R. Morency, J. Waite, and V. Schwanda. 1990. Ergonomic improvement in a Garr-tack sewing job: A case study. In *Advances in Industrial Ergonomics and Safety II,* ed. Biman Das. London: Taylor & Francis.

CHAPTER 2

Alexander, D. C. 1986. *The Practice and Management of Industrial Ergonomics.* Englewood Cliffs, NJ: Prentice–Hall.

Chaffin, D., and G. Anderson. 1984. *Occupational Biomechanics.* New York: John Wiley and Sons.

Grandjean, E. 1988. *Fitting the Task to the Man,* 4th edition. London: Taylor & Francis.

Human Factors Section, Eastman Kodak Company. 1983. *Ergonomic Design for People at Work,* Volumes I and II. New York: Van Nostrand Reinhold.

Kantowitz, B., and R. Sorkin. 1983. *Human Factors.* New York: John Wiley and Sons.

Kroemer, K. H. E., H. J. Kroemer, and K. E. Kroemer-Elbert. 1990. *Engineering Physiology,* 2nd edition. New York: Van Nostrand Reinhold.

McCormick, E., and M. Sanders. 1989. *Human Factors in Engineering and Design,* 6th edition. New York: McGraw–Hill.

Putz-Anderson, V. 1988. *Cumulative Trauma Disorders.* London: Taylor & Francis.

Woodson, W. E. 1981. *Human Factors Design Handbook.* New York: McGraw–Hill.

CHAPTER 3

Schneiderman, B. 1987. *Designing the User Interface.* Reading, MA: Addison–Wesley Publishing Company.

CHAPTER 4

Gray, H. 1896. *Anatomy, Descriptive and Surgical,* 13th edition. Philadelphia: Lea Brothers.

Ramazzini, B. 1713. *Diseases of Workers* (English translation Wilmer C. Wright, 1964). New York: Hafner.

Silverstein, B. A. 1985. *The Prevalence of Upper Extremity Disorders in Industry.* Ann Arbor: Center for Ergonomics, University of Michigan.

CHAPTER 5

Crosby, P. B. 1979. *Quality Is Free.* New York: McGraw–Hill.

Deming, W. E. 1986. *Out of the Crisis.* Cambridge, MA: MIT Center for Advanced Engineering Study.

Goldratt, E., and J. Cox. 1984. *The Goal.* Croton-on-Hudson, NY: North River Press.

Taylor, F. W. 1911. *The Principles of Scientific Management.* New York: Harper.

Tichauer, E. R. 1978. *The Biomechanical Basis of Ergonomics.* New York: John Wiley and Sons.

CHAPTER 6

MacLeod, Dan. 1984. Why Sweden has better working conditions than the U.S. *Working Life in Sweden,* Series No. 28. New York: Swedish Information Service.

CHAPTER 7

Arndt, Robert. 1983. Carpal tunnel syndrome in postal workers. Paper presented at the Occupational Cumulative Trauma Disorders of the Upper Extremities Conference, Ann Arbor, University of Michigan.

Arndt, Robert. 1987. Job pace, stress and cumulative trauma disorders. *Journal of Hand Surgery* 12A(5 pt 2):866–869.

Bernard, Bruce, Steven L. Sauter, Lawrence J. Fine, Martin R. Peterson, and Thomas R. Hales. 1992. Psychosocial and work organization risk factors for cumulative trauma disorders in the hands and wrists of newspaper employees. *Scandinavian Journal of Work Environment Health* 18(Suppl. 2):119–120.

Bigos, S., et al. 1991. A prospective study of work perceptions and psychosocial factors affecting the report of back injury. *Spine* 19(1):1–6.

Bongers, Paulien, Cornelius R. de Winter, Michiel A. J. Kompier, and Vincent

H. Hildebrandt. 1993. Psychosocial factors at work and musculoskeletal disease. *Scandinavian Journal of Work Environment Health* 19:297–312.

Cergol, S. 1991. Repetitive motion injury. *NTID Focus* Winter/Spring:11–16.

Dimberg, Lennart. 1991. *Symptoms from the Neck, Shoulders and Arms in an Industrial Population and Some Related Problems.* Göteborg, Sweden: Department of Orthopaedics, University of Göteborg.

Flor, H., D. C. Turk, and B. Nirbaumer. 1985. Assessment of stress-related psychophysiological reactions in chronic back pain patients. *Journal of Consulting and Clinical Psychology* 53:354–364.

Frankenhäuser, Marianne. 1981. Coping with stress at work. *International Journal of Health Services* 11(4):491–510.

Glass, D. C., and J. E. Singer. 1972. *Urban Stress: Experiments in Noise and Social Stresses.* New York: Academic Press.

Harris, Marvin. 1978. *Cannibals and Kings—The Origins of Cultures.* New York: Vintage Books.

Leino, Päivi. 1989. Symptoms of stress predict musculoskeletal disorders. *Journal of Epidemiology and Community Health* 43:293–300.

Linton, S. J., and K. Kamwendo. 1989. Risk factors in the psychological work environment for neck and shoulder pain in secretaries. *Journal of Occupational Medicine* 31(7):609–613.

MacLeod, Dan. 1988. Workplace and Personal Stresses and Insurance-Related Losses Due to Accidents and Injuries. Unpublished master's thesis, University of Minnesota Graduate School.

MacLeod, Dan. 1992. Work organization and cumulative trauma disorders. Draft report to the American National Standards Institute Z-365 Committee on Cumulative Trauma Disorders, Chicago.

Magnusson, M., et al. 1990. The loads on the lumbar spine during work at an assembly line. *Spine* 15(8):774–779.

Middlestadt, S. E., and M. Fishbein. 1988. Health and occupational correlates of perceived occupational stress in symphony orchestra musicians. *Journal of Occupational Medicine* 30(9):687–692.

NIOSH. 1990. Health Hazard Evaluation Report, HETA 89–250–2046.

NIOSH. 1992a. Health Hazard Evaluation Report, HETA 89–299–2230.

NIOSH. 1992b. Health Hazard Evaluation Report, HETA 90–013–2277.

Northwestern National Life. 1992. *Employee Burnout: Causes and Cures.* Minneapolis: Northwestern National Life Insurance Company.

Report on a World Health Organization Meeting. 1989. Work with visual display terminals: psychosocial aspects and health. *Journal of Occupational Medicine* 31(12):957–968.

Ryan G., B. Hage, and M. Bampton. 1987. Postural factors, work organization and musculoskeletal symptoms. In *Musculoskeletal Disorders at Work,* ed. P. Buckle. London: Taylor & Francis.

Smith, Michael J., Pascale Sainfort, Katherine Rogers, and David LeGrande.

1990. *Electronic Performance Monitoring and Job Stress in Telecommunications Jobs*. Madison: Department of Industrial Engineering, University of Wisconsin.

Theorell, Töres, Karin Harms-Ringdahl, Gunnel Ahlberg-Hultén, and Bibbi Westin. 1991. Psychosocial job factors and symptoms from the locomotor system—a multicausal analysis. *Scandinavian Journal of Rehabilitation Medicine* 23:165–173.

Udy, Stanley H., Jr. 1959. *Organization of Work—A Comparative Analysis of Production Among Nonindustrial Peoples*. New Haven: Yale University Press.

CHAPTER 9

MacLeod, Dan, Eric Kennedy, and Wayne Adams. 1992. *Setting Up an Ergonomics Program*. Minneapolis: Clayton ErgoTech.

MacLeod, Dan, Eric Kennedy, and Wayne Adams. 1993. *The Ergonomics Process*. Minneapolis: Clayton ErgoTech.

INDEX

Page numbers followed by f and t represent Figures and Tables respectively

Date Due